7th International Conference on Mechanical, Automotive and Materials Engineering 2019

AA007419

IOP Conference Series: Materials Science and Engineering
Volume 831

Melbourne, Australia
8 - 10 December 2019

ISBN: 978-1-7138-2678-1
ISSN: 1757-8981

Printed from e-media with permission by:

Curran Associates, Inc.
57 Morehouse Lane
Red Hook, NY 12571

Some format issues inherent in the e-media version may also appear in this print version.

Institute of Physics
Dirac House, Temple Back
Bristol BS1 6BE UK

Phone: 44 1 17 929 7481
Fax: 44 1 17 920 0979

techtracking@iop.org

Additional copies of this publication are available from:

Curran Associates, Inc.
57 Morehouse Lane
Red Hook, NY 12571 USA
Phone: 845-758-0400
Fax: 845-758-2633
Email: curran@proceedings.com
Web: www.proceedings.com

TABLE OF CONTENTS

MATERIAL DESIGN AND MOLDING PROCESS

MECHANICAL ENGINEERING AND ADVANCED MANUFACTURING SYSTEM

IOP Conf. Series: Materials Science and Engineering **831** (2020) 011001 doi:10.1088/1757-899X/831/1/011001

PREFACE

The proceedings contain the scholarly papers presented in two reputable joint conferences, the 7th International Conference on Mechanical, Automotive and Materials Engineering (CMAME 2019) with its workshop the 2nd International Conference on Progress in Mechanical and Aerospace Engineering (PMAE 2019). Both conferences have been held in different parts of the world for some years, as indicated by the number sequence. This year (2019), the conferences are held in Melbourne, Australia.

The conferences have invited keynote speakers who are professors from renowned universities in Australia, England and China. Together with the assistance of the International Technical Committee (members from universities and research organisations around the world), the conference attracted many submissions of high quality papers. Hence, the papers in these proceedings have passed the vigorous review process and amended to the satisfaction of reviewers and have undergone critiques in a face-to-face presentation with expert audience. It is on this basis that these proceedings are good references to future research investigations and publications.

The proceedings have four sections. Section 1 (Engineering Material Design and Analysis) has papers on different treatment of materials or related processes. Study of materials properties is essential to develop efficient machining and formation processes for cost effective manufacturing. The papers are good reflection of the ultimate goals of materials engineering research.

Section 2 (Mechanical Engineering and Intelligent Robots) papers focus on robotics and kinematics. Applications range from inspection to localisation and sensors. The broad application areas of robots in industry are highlighted in the short conference version of the papers. It is interesting to note that authors of this section come from countries such as Australia, China and South Korea

Section 3 (Applied Mechanics and Engineering) papers are related to more fundamental treatment of fluid mechanics, mechanical design, drive systems and residual strain problems. These studies have no immediate industry applications but the theoretical knowledge outcomes form the more foundation level research will contribute to future development of products requiring these technologies.

Section 4 (Advanced Manufacturing Systems and Technologies) papers are more application oriented investigating real-life industry problems in ship building, machining discrepancies, automotive crash, product life cycle analysis, energy consumption and non-linear control systems. These applications are researched in universities from 4 countries around the world, illustrating the benefits of a face-to-face conference allowing exchange of ideas and latest research findings.

The variety of research topics presented in the conference and novelty exhibited in the papers published in these proceedings once again demonstrated the value of the two conferences jointly organised enabling researchers of different nationalities and different engineering fields to participate in a live research conversation. I would like to thank the conference committee, the conference secretariat, keynote speakers and all participants for their hard work making this conference a success as well as publishing a valuable knowledge literature contributing to state-of-the-art developments in mechanical, automotive, materials and aerospace engineering.

John P.T. Mo
Professor of Manufacturing Engineering
BSc(Engg), MSc(Engg), PhD(Loughborough), FIEAust CPEng NER APEC Engineer IntPE(Aus), CEng, FIMechE, MHKIE, MFIE
School of Engineering
RMIT University

December 2019, Melbourne, Australia

IOP Conf. Series: Materials Science and Engineering **831** (2020) 011002 doi:10.1088/1757-899X/831/1/011002

CONFERENCE COMMITTEES

Conference Committee Chairs

John Mo, Royal Melbourne Institute of Technology, Australia

Yonggang Zhu, Harbin Institute of Technology - Shenzhen, China

Program Committee Chairs

Yu-Liang Chen, National Defense University, Taiwan, China

Xingjian Jing, The Hong Kong Polytechnic University, Hong Kong

International Technical Committees

Alokesh Pramanik, Curtin University, Australia

BELAL F. YOUSIF, The University of Southern Queensland, Australia

Effendi Mohamad, Universiti Teknikal Malaysia Melaka, Malaysia

Guangning Li, Northwestern Polytechnical University, China

Hassan Ghadbeigi, The University of Sheffield, UK

Jingjing Xu, Shanghai University, China

Jingwei Zhao, University of Wollongong, Australia

Kheng-Lim Goh, Newcastle University, UK

Mali ABDOLLAHIAN, Royal Melbourne Institute of Technology, Australia

Mohd Rizal Salleh, Universiti Teknikal Malaysia Melaka, Malaysia

Muhamad Arfauz A Rahman, Universiti Teknikal Malaysia Melaka, Malaysia

Pooya Sareh, Imperial College London, UK

Prodip Das, Newcastle University, UK

Qun Wei, Keimyung University, South Korea

Thanakom Soontornchainacksaeng, King Mongkut's University of Technology North Bangkok, Thailand

Timotius Pasang, Auckland University of Technology, New Zealand

Vinod Kushvaha, Indian Institute of Technology Jammu, India

Wong Shaw Voon, Universiti Putra Malaysia, Malaysia

Xiangsheng Gao, Beijing University of Technology, China

IOP Conf. Series: Materials Science and Engineering **831** (2020) 011002 doi:10.1088/1757-899X/831/1/011002

Yat Wong, Swinburne University of Technology, Australia

Zengtao Chen, University of Alberta, Canada

Shamsul Jamel, Universiti Teknologi MARA, Malaysia

Shiqi Zheng, China University of Geosciences, China

Suvash C. Saha, University of Technology Sydney (UTS), Australia

IOP Conf. Series: Materials Science and Engineering **831** (2020) 011003 doi:10.1088/1757-899X/831/1/011003

Peer review statement

All papers published in this volume of *IOP Conference Series: Materials Science and Engineering* have been peer reviewed through processes administered by the proceedings Editors. Reviews were conducted by expert referees to the professional and scientific standards expected of a proceedings journal published by IOP Publishing.

IOP Conf. Series: Materials Science and Engineering **831** (2020) 012001 doi:10.1088/1757-899X/831/1/012001

A Kind of Triple-negative Elastic Metamaterial

X Wang, S Wan*

Department of Transportation, Southeast University, Nanjing210096, China

wangxiao1133@126.com, lanyu421@163.com

Abstract. In this paper, a kind of triple-negative elastic metamaterial (EMM) is proposed, which means it possesses negative equivalent mass density, negative equivalent bulk modulus, negative equivalent shear modulus in the same frequency range simultaneously, and can realize negative refraction of both transverse and longitudinal waves. Furthermore, the influence of rectangular oscillators and semi elliptical oscillators on the band structure is analysed, and we find that the unit cell with semi elliptical oscillators has a much wider band gap, which has a potential in supressing vibration at a wide frequency range.

1. Introduction

Double-negative elastic metamaterials (EMMs) were proposed by analogy with double-negative electromagnetic metamaterials. In 1967, Russian physicist Victor Veselago first predicted a kind of medium with negative dielectric constant (ε) and negative permeability (μ) simultaneously in an article [1]. He also indicated that the medium had a negative refractive index. John Pendry, a British physicist, first designed structural materials with negative equivalent ε and negative equivalent μ in 1996, which made Veselago's predictions come true [2]. At the beginning of 2000, Sheng Ping research group of the University of Hong Kong validated the local resonance of elastic waves in three-dimensional coated spherical arrays by numerical simulation and experiments [3]. This work layed a foundation for the analogy between electromagnetic metamaterials and acoustic metamaterials. A EMM is a well-designed and processed artificial composite structure with special characteristics. It can realize wave focusing [4], anisotropy [5], and refractive index changing with position [6]. There are far-reaching and broad application prospects in engineering fields, such as sub-wavelength waveguide [7], low frequency vibration reduction [8], seismic wave stealth of buildings [9], etc. The design of EMMs is more challenging than that of electromagnetic metamaterials [10]. For now, the mechanism of double-negative EMMs is clear [11]. Negative equivalent mass density (ρ_{eff}) is produced by dipole resonance. Negative equivalent bulk modulus (E_{eff}) is produced by monopole resonance. Negative equivalent shear modulus (G_{eff}) is produced by quadrupole resonance. In addition, many kinds of double-negative EMMs which can realize double negative parameters simultaneously in the same frequency range has been designed [12] [13]. The refractive index of longitudinal (P) waves is negative when ρ_{eff} and E_{eff} are both negative simultaneously, and the refractive index of transverse (S) waves is negative when ρ_{eff} and G_{eff} are both negative simultaneously. Negative refraction facilitates the propagation control of elastic waves. Moreover, triple-negative EMMs which mean ρ_{eff}, E_{eff} and G_{eff} are negative simultaneously in the same frequency range may have significance in elastic wave propagation control. Double-negative EMMs mainly could be used in acoustic devices, because sound waves are P waves usually. However, in mechanical engineering and civil engineering, etc., there are both P waves and S waves in structural members. So, it is meaningful to study triple-negative EMMs. In this paper, a kind of triple-negative EMMs that can realize three negative parameters in the same frequency range is proposed, which may provide a reference for the design of triple-negative EMMs.

2. Modelling

The two-dimensional unit cell of the triple-negative EMMs we proposed is shown in figure 1. The lattice constant is $a = 6\text{mm}$. A round hole with radius R and four oval holes with major axis D_1 and short axis D_2 are excavated on the square foam matrix. Half of every elliptical hole is filled with tungsten as a scatterer, and the distance from the vertex of the ellipse to the center of the circle is D_4. The blank parts are vacuums. The four tungsten blocks are centrosymmetric, which make it easier to form multiple resonance. The material parameters of foam are ρ=115kg/m^3, λ=6$\times$10^{6}N/m^2, μ=3$\times$10^{4}N/m^2, and the material parameters of tungsten are ρ=19300kg/m^3, λ=1.974$\times$10^{11}N/m^2, μ=1.513$\times$10^{11}N/m^2, where λ and μ are the lame constants and ρ is the mass density. The band structure and eigenmodes of the unit cell are calculated by COMSOL.

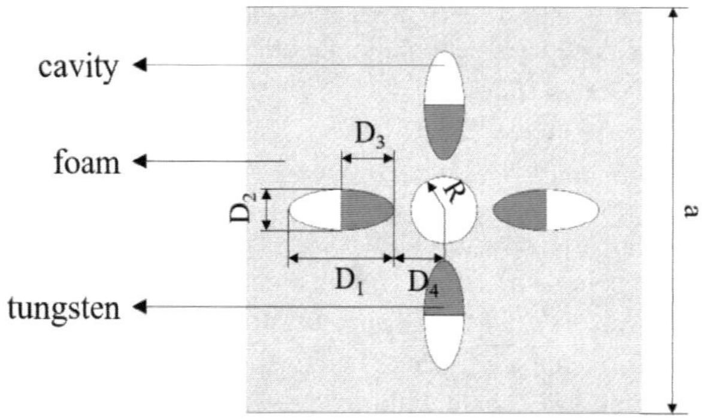

Figure 1. Unit cell design drawing of the triple-negative metamaterial. Here, $a = 6\text{mm}$, $R=1\text{mm}$, $D_1 = 1.6\text{mm}$, $D_2 = 0.6\text{mm}$, $D_3 = 0.8\text{mm}$, $D_4 = 0.3\text{mm}$

3. Analysis

The band structure of the calculated ΓX direction is shown in figure 2. The 5th band and 6th band are negative, and the negative equivalent parameters of the unit cell appear on these two bands. The 5th negative band is a transverse wave band propagating S waves only and the 6th negative band is longitudinal wave band propagating P waves only, which will be proved later. For the convenience of observation, these two negative bands are enlarged as shown in figure 3. The 5th band curve is represented by line A with a starting point A1, and the 6th band curve is represented by line B with a starting point B1. The 5th negative band starts at 8097.9Hz and ends at 7668.1Hz. The 6th negative band starts at 8154.8Hz and ends at 7812.9Hz. A2 and B2 are the points on line A and line B separately.

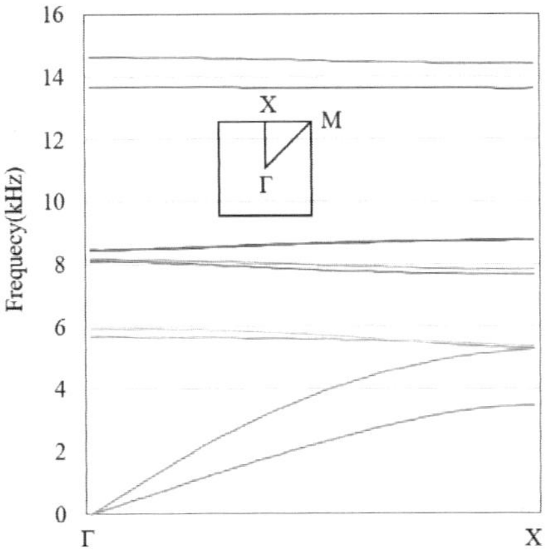

Figure 2. Band structure of the unit cell in figure 1 along the ΓX direction

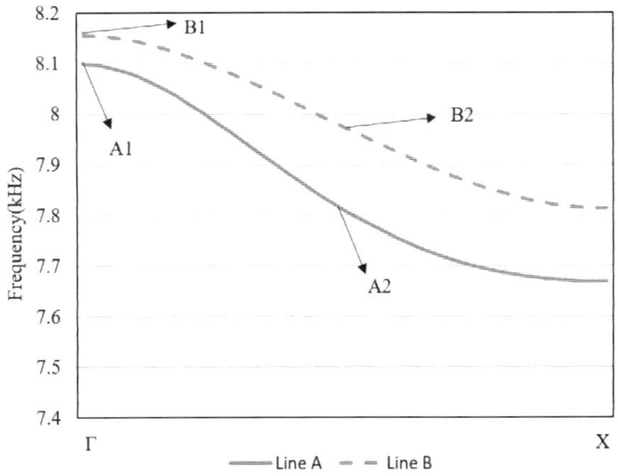

Figure 3. Zoomed band structure of the 5th and 6th band in figure 2

The mode of A1, as shown in figure 4 (a), is quadrupole resonance formed by the rotation of four metal tungsten blocks. The quadrupole resonance results in negative equivalent shear modulus, which has an effect on the propagation of P waves. As shown in figure 4 (b), the mode of B1 is monopole resonance, which is formed by four tungsten blocks moving away from the center at the same time. Monopole resonance results in negative equivalent bulk modulus, which has an effect on the propagation of S waves. The mode of B2 is shown in figure 4 (c). The mode is composed of monopole and dipole resonance coupling, which can realize negative ρ_{eff} and E_{eff} simultaneously. The mode of A2 is shown in figure 4 (d). The mode is composed of quadrupole and dipole resonance coupling, which can realize negative ρ_{eff} and G_{eff} simultaneously.

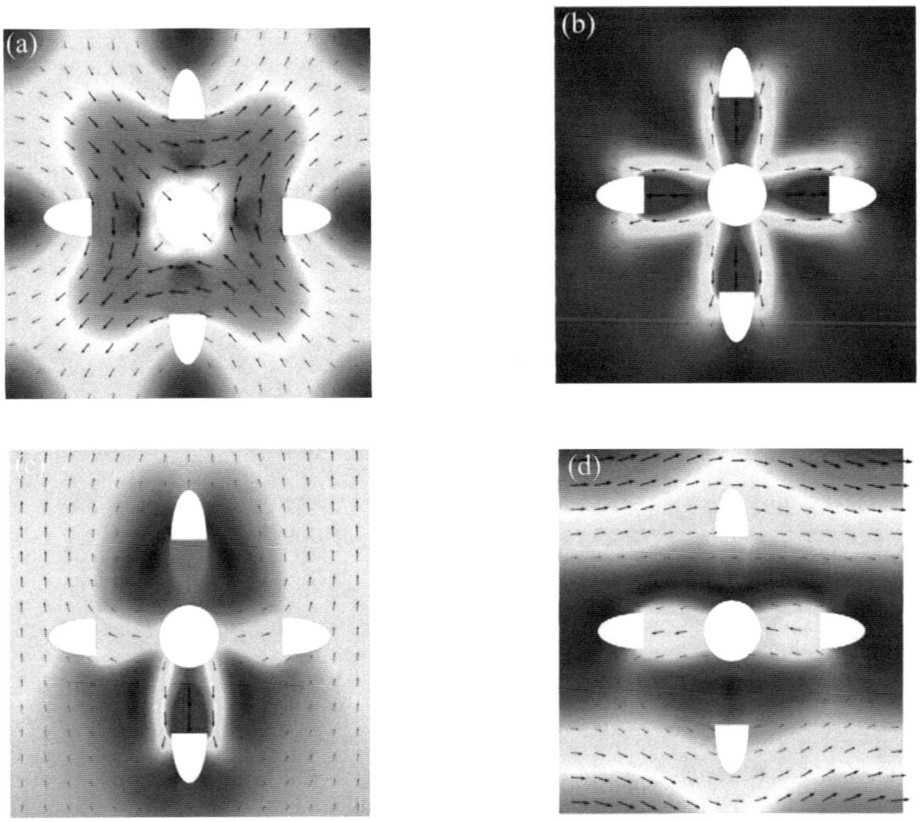

IOP Conf. Series: Materials Science and Engineering **831** (2020) 012001 doi:10.1088/1757-899X/831/1/012001

Figure 4. (a)~(d) Eigenfields of point A_1, B_1, B_2, A_2 respectively.
Arrows represent field directions, and colour red represents large amplitude, and colour blue represents small amplitude.
(e) Sketch map of hybridization of the monopolar and dipolar mode shown in (c).
(f) Sketch Map of hybridization of the quadrupolar and dipolar mode shown in (d).

In order to further prove that band A is a transverse wave band and band B is a longitudinal wave band, it is necessary to study the response of the two bands to P and S waves. The established verification model is shown in figure 5, with 10 cells connected to form a metamaterial belt. The source line on the left end of the belt is the signal input end, and the probe line is the signal receiving end. Perfectly Matched Layers （PMLs）should be set at both ends of the model to prevent the interference of reflected waves, and periodic conditions should be set to the upper and lower sides of the model. If P waves and S waves are input into the belt separately, we can see that the mode of S wave incidence is quadrupole resonance, and the mode of P wave incidence is monopole resonance, which is consistent with the previous analysis. Besides, in this case, the transmission of the model is calculated, as is shown in figure 6. It can be seen that the S wave transmissibility is higher in the frequency range of the 5th band and the P wave transmissibility is higher in the frequency range of the 6th band. Therefore, it is verified that the 5th band is a transverse wave band and the 6th band is a longitudinal wave band. The transmissibility of S waves and P waves is both very high in the overlapping frequency range of the two bands, so the negative ρ_{eff}, negative E_{eff} and negative G_{eff} can be realized simultaneously in this overlapping frequency range.

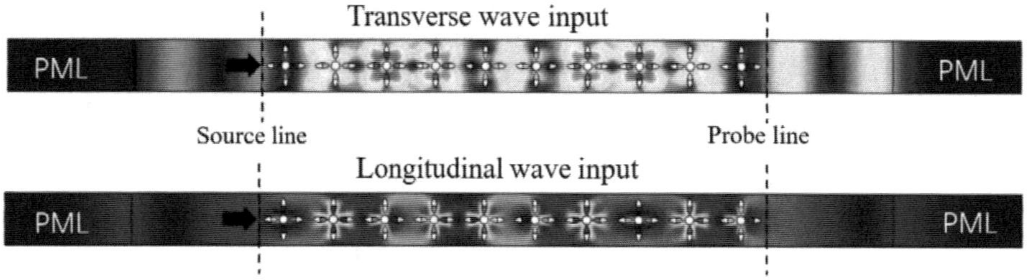

Figure 5. Model of a EMM belt composed of ten unit cells.

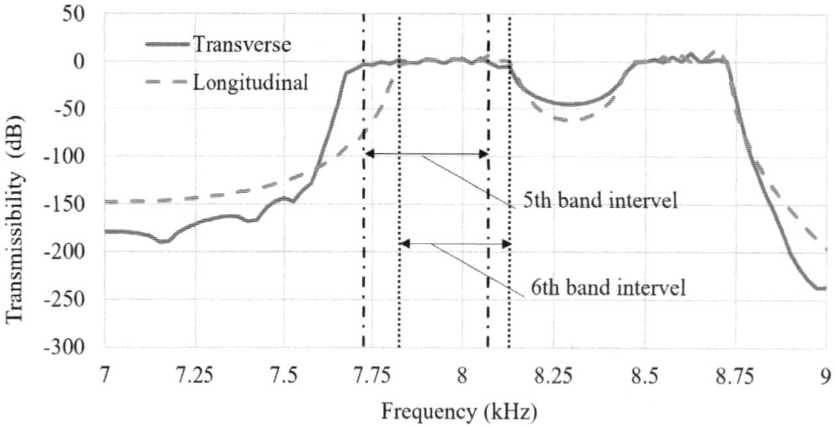

Figure 6. Transmission of S wave and P wave incident on the model in figure 5.
The frequency intervals of the 5th band and 6th band are shown in this figure.

IOP Conf. Series: Materials Science and Engineering **831** (2020) 012001 doi:10.1088/1757-899X/831/1/012001

4. Comparison and Discussion

The unit cell proposed in this paper has a wide band gap, which is a peculiarity. Figure 7 shows the structure of a unit cell proposed by Wu et al [11]. It has rectangular oscillators and also can realize three negative parameters in the same frequency range simultaneously. Its band structure is shown in figure 8. The width of the first band gap is almost the same as that of the cell proposed in this paper, and its width is about 2KHz. The second band gap width is about 3kHz, while the second band gap width of the unit cell proposed in this paper is about 4.5khz. In order to make the unit cell with rectangular oscillators produces a wider band gap, the mass of the tungsten block can be reduced by half (L reduced from 1mm to 0.5mm), and the band structure is shown in figure 9. However, it can be seen that there is not any overlapping part of the 5th and 6th band, which means the unit cell loses the characteristic of three negative parameters in the same band simultaneously. The combination of wide band gaps and negative refraction can be used for vibration reduction and isolation. Band gaps prevent elastic waves from propagating at certain frequency ranges, and for the elastic waves that can't be stopped by band gaps, negative refraction can facilitate making the wave propagation path change so that the waves don't propagate to the target. Based on this principle, functional devices that can suppress vibration at a wide frequency range can be fabricated， such as functionally graded materials and directional invisibility cloaks.

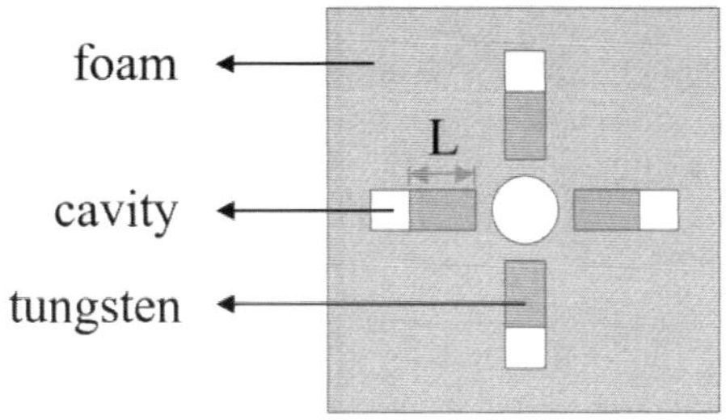

Figure 7. A unit cell with rectangular oscillators that can can realize three negative parameters in the same frequency range simultaneously

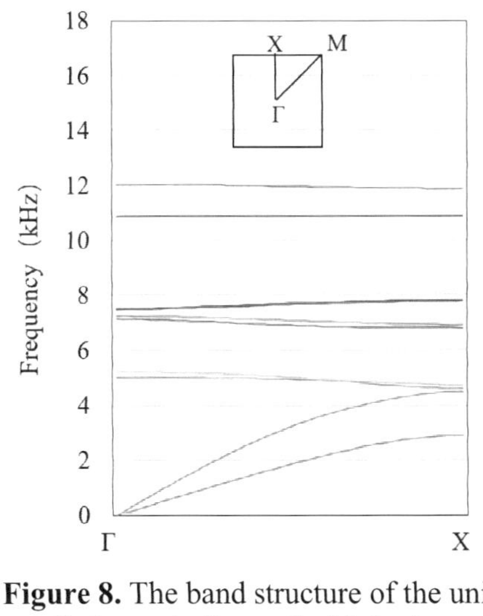

Figure 8. The band structure of the unit cell shown in figure 7 with L=1mm

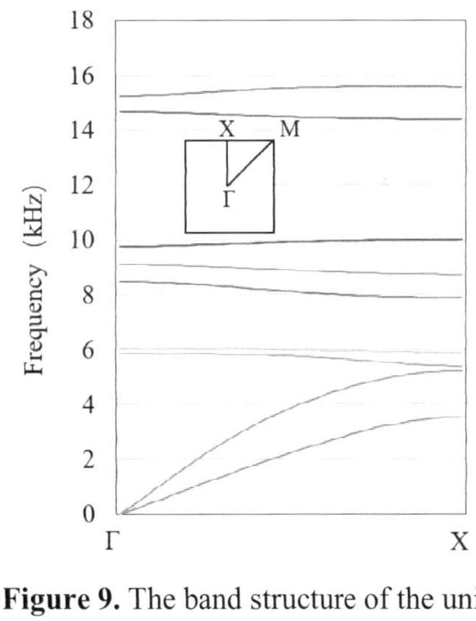

Figure 9. The band structure of the unit cell shown in figure 7 with L=0.5mm

5. Conclusion

In conclusion, we have proposed a novel triple-negative EMM with a wide band gap, which may enlighten the design or optimization of triple-negative EMMs for vibration isolation and reduction. Based on studying the eigenmodes and transmission of the triple-negative EMM, we have proved that it can achieve negative equivalent ρ_{eff} E_{eff} and G_{eff} at the same frequency range simultaneously, which can facilitate the propagation control of both P and S waves. We also find that it has a wide band gap, so we put forward the idea of using both passbands and band gaps to isolate vibration.

6. References

[1] Veselago V G 1968 *Sov. Phys. Usp.* **10** 509
[2] Pendry J B, Holden A J, Stewart W J 1996 *Phys. Rev. Lett.* **76** 4773-4776
[3] Liu Z Y, Zhang X X, Mao Y W, Zhu, Y Y, Yang Z Y, Chan C T, Sheng P 2000. *Science* **289** 1734-1736
[4] Yang S X, Page J H, Liu Z Y, Cowan M L, Chan C T, Sheng P 2004 *Phys. Rev. Lett.* **93** 1-4
[5] Dong H W, Zhao S D, Wang Y S, Zhang C Z 2017 J. *Mech. Phys. Solids* **105** 54-80
[6] Brun M, Guenneau S, Movchan A B *Appl. Phys. Lett.* **2009** 1-3
[7] Jin Y B, Fernez N, Pennec Y, Bonello B, Moiseyenko R P, Hemon S, Pan Y D, Djafari-Rouhani B 2016 Phys. Rev. B **93** 1-8
[8] Zhu R, Liu X N, Hu G K, Sun C T, Huang G L 2014 *J. Sound Vib.* **33** 2759-2773
[9] Farhat M, Guenneau S, Enoch S, Movchan A B 2009 *Phys. Rev. B* **79** 1-4
[10] Christensen J, Kadic M, Wegener M, Kraft O, Wegener M 2015 *MRS Commun.* **5** 453-462
[11] Wu J E, Wang X Y, Tang B, He Z J, Deng, K, Zhao H P 2017 *AIP Adv.* **7** 1-6
[12] Wang W, Bonello B, Djafari-Rouhani B, Pennec Y, Zhao J F 2018 *Phys. Rev. Appl.* **10** 1-9
[13] Kumar S, Bhushan P, Prakash O, Bhattacharya S 2018 *Appl. Phys. Lett.* **112** 1-14

IOP Conf. Series: Materials Science and Engineering **831** (2020) 012002 doi:10.1088/1757-899X/831/1/012002

Interaction Effect of Machining Parameters on Material Removal Rate in the Machining of AA6061–T6 Using Minimum Quantity Lubrication Conditions

M.S. Najiha[1], M.M. Rahman[2,*], K. Kadirgama[2], and D. Ramasamy[2]

[1]International Affairs and training Division Pakistan Atomic Energy Commission, Karachi, Pakistan
[2]Faculty of Mechanical and Automotive Engineering Technology, Universiti Malaysia Pahang, 26600 Pekan, Pahang, Malaysia

mustafizur@ump.edu.my

Abstract. The objectives of this study are to investigate the interaction effect of machining parameters on material removal rate in end milling of aluminium alloy AA6061-T6 with conventional MQL techniques. Uncoated tungsten carbide (WC-Co 6.0%) and PVD TiAlN and TiAlN+TiN coated carbide cutting tools are considered using 23.4-54.0 ml/hr flow rate of commercial mineral oil for MQL machining with different combinations of input cutting parameters. Response surface methodology with central composite design approach is used for the design of experiments. Second-order mathematical models are developed for machining performance measures with different cooling conditions and validated statistically. The developed models show good agreement (< 5 % error) with the experimental results. The metal cutting performance of the TiAlN coated tools relative to uncoated and TiN coated inserts is better at all combinations of input cutting parameters. Hence the superior performance of TiAlN coating makes it more suitable for use with MQL. In case of material removal rate, all the tools show similar behaviour in all the measurements with the depth of cut as the most significant parameter followed by feed rate. Interaction of feed rate and depth of cut is most effective.

1. Introduction

As the manufacturing world is in continuous pursuit of investigating the methods in order to increase the process performance and to reduce the production costs, in addition to the growing environmental concerns [1], minimum quantity lubrication process can offer the near-term solution to the problem. Minimum quantity lubrication (MQL) has demonstrated as a successful near-dry machining technique as well as a globally-acknowledged option compared to completely dry and wet cutting conditions from the perspective of cost, ecological, human health issues and machining process performance [2]. The significant challenges faced by the manufacturing industry are the improved quality, enhanced productivity as well as economic production [3-4]. These challenges are addressed by increasing the material removal rate for enhanced productivity, surface quality and surface integrity as well as longer tool life with consistence performance [5]. Cutting fluids are considered essential for machining operations in order to perform lubrication, cooling and chip flushing. These functions of cutting fluids in machining processes are continually being reviewed due to cost pressures together with growing global concerns related to occupational and environmental consciousness [6-8] and the want for increased employee satisfaction through healthier environment and cleaner work areas [5]. The

conventional method of application of cooling and lubrication in machining processes involve excessive use of cutting fluids.
In MQL, the cutting fluid flow rate varies from a minimum of 10 ml/hr to a maximum of 100 ml/hr [8]. Some researchers have given different ranges of lubricant flow rate in MQL, such as 2 to 500 ml/hr [9]. However, most industrial applications consume MQL in the range of 10 to 100 ml/hr [[9]. High pressure of the fluid penetrates through the rake face of the tool hence lowering the temperature of the critical zone. It was observed that oil-water MQL had better cooling characteristics while pure oil MQL had better lubrication capacity. Navid [10] analysed the results from dry and MQL machining of Monel K-500 in order to investigate the influence of cutting speed, depth of cut and feed rate on surface roughness and cutting forces. MQL yielded a reduction in surface roughness by 38 % and cutting force by 59 %. The vegetable-based lubricant used to conduct wear analysis experiments during end milling AISI 1018 steel with varying feed rate and cutting speed at a fixed depth of cut [11]. Solid carbide cutting tool was used with varying cutting speed and feed rate has a constant depth of cut. The proper selection of the cutting parameters might result in possible to obtain higher tool life. The objective of this study to investigate the interaction effect of machining parameters on the material removal rate in the machining of aluminium alloy AA6062-T6 for minimum quantity lubrication technique.

2. Materials and Methods

Aluminium alloy AA6061-T6 is selected as a workpiece material due to excellent mechanical and corrosion resistance [12]. Conflicting views about the cooling conditions for the alloy are observed in the literature [13]. Workpiece dimensions are 100 × 100 × 30 mm. Alloy composition is recorded using a spectrometer (Foundary-Master type, Oxford Instruments, Inc.) for three random samples at three different places each and weight % obtained is an average of the nine samples. The conventional (oil-based) MQL cutting environments are considered in the study. Vegetable oil-based coolant (Coolube 2210, UNIST, Inc.) is used for conventional MQL machining. A two-flute end mill of diameter 12 mm MaxiMill shoulder design cutter (CERATIZIT: C211.12.R.02-07-A-20) is used. Three types of different milling inserts including uncoated WC-Co 6.0%, TiAN coating and TiAlN+TiN coating by PVD process are selected. The facet corner radius for uncoated tungsten carbide insert is 0.8 μm while for coated inserts facet corner radius is 0.4 μm. Cutting edge width of uncoated tungsten carbide inset is 0.1 mm more than coated ones. All other dimensions are similar. Two-flute end mill cutter is selected to minimise the clogging of adhered material and chips into the end mill flutes. Length of the end mill flute is 8.0 mm. MQL is delivered to the cutting zone through the UNIST Uni-MAX Coolubricator system. The pulse generator produces repetitive cycles of the metering pump. A 1-drop meter output 6-nozzles configuration is used. The system is provided with six positive displacement oil metering pumps (0.2-1.00 ml/stroke) with adjustable stroke and output. MQL flow rate is adjusted by setting the stroke length as 10 while the flow rate is adjusted by indexing the flow rate according to counter-clockwise turnings. Material removal rate (MRR) is a critical machining control factor from the viewpoint of productivity. The material removal rate is calculated in mm^3/min. Schimadzu (4-decimal places resolution) digital balance is used to measure the mass of workpiece before and after the machining for material removal rate. Mass of the workpiece is measured after machining a slot over the entire length of a workpiece according to the design of an experiment. Machining time for each sample has been calculated accordingly. The machining is performed with machine doors closed. Response surface methodology is used to determine the optimum operating conditions. Theoretical formulae can be used to estimate material removal rate in end milling however in order to estimate actual material removal rate with material density included in analysis along with the machining conditions effects, reduction in volume per unit time is used for more actual results. Hence, empirical relationships are necessary to relate to the material removal rate and input machining parameters. Levels of independent cutting parameters in the experimental design for material removal rate under conventional MQL is presented in Table 1. Total 78 (26 × 3) experiments result for conventional MQL conditions with two experiments repeated to increase the model accuracy and repeatability.

Table 1. Levels assigned to the input cutting parameters for conventional MQL cooling conditions.

Input cutting parameter	Levels of input cutting parameters				
	Lowest	Low	Centre	High	Highest
	Conventional MQL conditions				
Cutting speed (RPM)	5252.0	5300.0	5400.0	5500.0	5548.0
Feed rate (mm/min)	288.0	318.0	379.0	440.0	469.0
Depth of cut (mm)	0.37	1.00	2.00	3.00	3.63
MQL flow rate (ml/min)	0.39	0.48	0.65	0.83	1.00

3. Results and Discussion

3.1 Mathematical Modeling

The mathematical models in terms of material removal rate are developed using response surface method. Second-order mathematical models for different tools with various cooling environments are developed using RSM. Analysis of variance (ANOVA) is utilized to verify the adequacy of the experimental data. The estimated regression coefficients for the material removal rate with various cutting inserts are determined using experimental data. The ANOVA for MRR with uncoated tungsten carbide, TiAlN coated and TiAlN+TiN coated carbide tools are presented in Table 2. For conventional MQL machining, four squared terms (speed and speed, feed rate and feed rate, depth of cut and depth of cut and squared term of MQL flow rate), six interaction terms (speed and feed rate, speed and depth of cut, speed and MQL flow rate, feed rate and depth of cut, feed rate and MQL flow rate and interaction of depth of cut and MQL flow rate) and four linear terms are included in regression analysis. The regression *p*-value for different inserts is 0.000. Hence, regression analysis is significant. The interaction term of feed rate with the depth of cut shows a *p*-value of 0.000 <0.05; hence, the term is the most significant term in the analysis for all the inserts. For TiAlN coated insert in MQL machining, quadratic terms of depth of cut are also significant. It is observed that R-sq value for all the inserts is more than 98 % for various cooling conditions. Lack-of-fit for all the conditions is also insignificant i.e. greater than 0.05 for all cases. Therefore, it is concluded that the experimental data for all inserts with various cooling conditions are adequate and suitable for further analysis.

Table 2. ANOVA analysis for different uncoated- tungsten carbide and coated carbide insert.

	p-value		
	uncoated tungsten carbide	Coaded TiAlN	Coated TiAlN+TiN
Regression	0.000	0.000	0.000
Linear	0.142	0.181	0.243
Square	0.109	0.093	0.001
Interaction	0.007	0.000	0.000
Lack-of-fit	0.219	0.088	0.054
R-Sq (%)	99.09	99.64	99.84

3.2 Effects of Input Parameters on Response Variables

Effects of input process parameters and different phenomena arising during machining on performance parameters against the experimental design are presented in the following section. Significance of input parameter is determined based on the difference in the mean of two groups of experimental design at high and low levels. The relative importance and rankings of main parameters and their interactions with respect to response variables are evaluated by determining the parameters effect size using DOE. Significance of input cutting parameters for material removal rate with conventional MQL is presented in Figure 1. Depth of cut is the most significant factor affecting material removal rate (61.06 -71.61 %) followed by feed rate (26.06 - 31.99 %). Theoretical formulae for material removal rate show that the response variable increases with increasing depth of cut and feed rate. This is due to more volume of chips generated with increasing depth of cut and increased material removed per revolution [14]. Depth of cut with larger intervals between the levels selected for the machining results in more significant effects of depth of cut on material removal rate in this study. Effects of speed on material removal rate are very insignificant. The reason is that the selected speed range is set at levels with very narrow intervals, hence the effects of speed are not so pronounced.

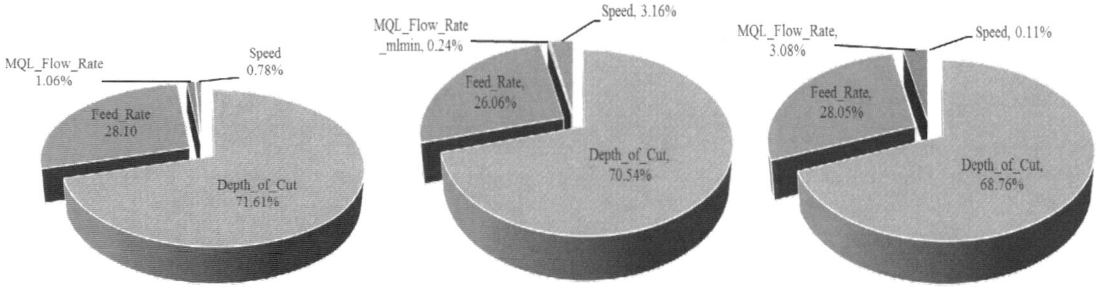

(a) uncoated tungsten carbide (b) TiAlN coated (c) TiAlN+TiN coated

Figure 1. Significance of input parameters for the material removal rate for different cutting tools with conventional MQL conditions

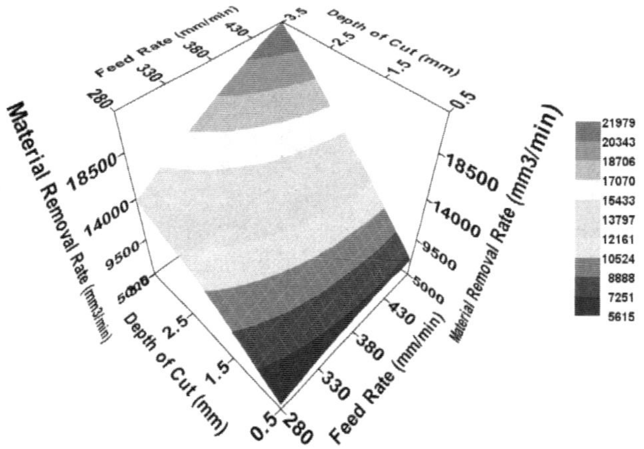

(a) Uncoated tungsten carbide insert

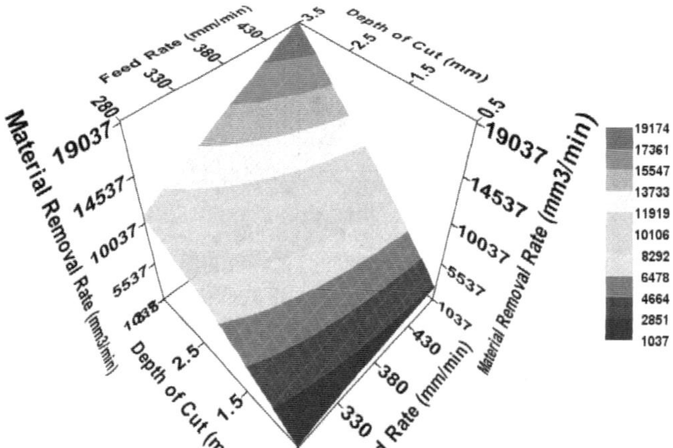

(b) TiAlN coated insert

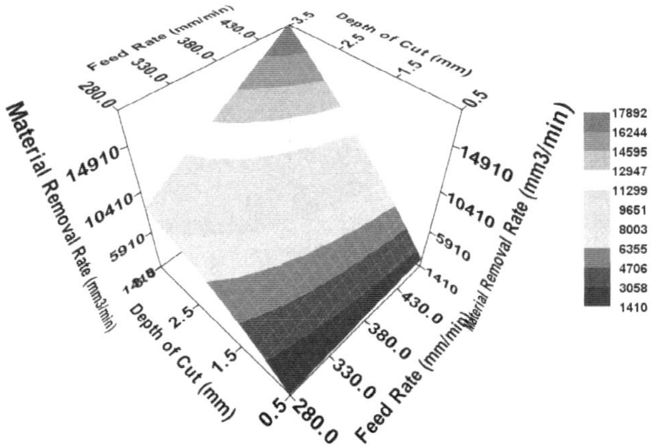

(c) TiAlN+TiN coated insert

Figure 2. Variations of material removal rate against the depth of cut and feed rate for conventional MDL conditions.

Figure 2 presents three-dimensional surface plots on the basis of response model of material removal rate against the depth of cut and feed rate for various inserts at a cutting speed of 5400 rpm and MQL flow rate of 0.65 ml/min. A linearly increasing relationship is observed for material removal rate with the depth of cut and feed rate. For all inserts used with different cooling conditions, the most significant parameters for the material removal rate are feed rate and depth of cut. Material removal rate increases with increasing depth of cut and feed rate but the effect of depth of cut is more significant. The maximum material removal rate is obtained at a maximum depth of cut and maximum feed rate due to the resulting maximum volume of chips and material removed. It is observed that variations in depth of cut more significantly affect the material removal rate as compared to feed rate. This is attributed to the larger volume of material removed on account of the higher depth of cut selected. All the tools in different cooling conditions show an average increasing rate of material removal rate per mm of the depth of cut when the feed rate is at its maximum. Hence a combination of higher depth of cut and higher feed rate is more effective for material removal rate for machining of AA6061 because of the higher material volume of chips produced with increasing depth of cut and feed rate. This conclusion is in complete accordance with the findings of Kuttolamadom et al. [15] for end milling of aluminium alloy 6061 T6. In order to increase the material removal rate, depth of cut and feed rate should be increased however the real challenge is to keep the surface roughness within the recommended range.

4. Conclusions

End milling experiments for aluminium alloy 6061 T6 are conducted in order to assess the performance of minimum quantity lubrication technique using uncoated and PVD coated cemented carbide tools. The PVD coatings used are single layer TiAlN and dual-layered TiAlN+TiN (TiN is the outer layer) coating on cemented carbide substrate. MQL machining was performed with uncoated carbide insert in order to assess the improvement in the insert performance. The central composite design approach of response surface methodology is used to design experiments. Second-order polynomials in terms of input cutting parameters including cutting speed, feed rate, depth of cut, MQL flow rate result as regression models from the experimental data to predict the machining performance measures in terms material removal rate. The models are validated statistically as well as experimentally. Regression models show good agreement with the experimental data. The metal cutting performance of the TiAlN coated tools relative to uncoated and TiN coated inserts is better at all combinations of input cutting parameters. In the case of TiAlN coated tools, not only the hot hardness of the tool plays a significant role in better performance at higher temperatures but the formation of the layer of Al_2O_3 also serves to increase the stability. Hence the superior performance of TiAlN coating makes it more suitable for use with MQL. All the tools show similar behaviour in all the measurements with the depth of cut as the most significant parameter followed by feed rate. Interaction of feed rate and depth of cut is also effective.

5. References

[1] D.F. Fratila: Proc. Romanian Acad. Series A, Vol. 14 (2013), pp. 56-63.
[2] B. Boswell, M.N. Islam, I.J. Davies, Y.R. Ginting, A.K. Ong: Int. J. Adv. Manuf. Technol. Vol. 92 (2017), pp. 321–340.
[3] M.S. Najiha, M.M. Rahman, K. Kadirgama, M.M. Noor and D. Ramasamy: Inter. J. Automot. Mech. Eng. Vol. 12(2015), pp. 3003-3017.
[4] M. Yogeswaran, K. Kadirgama, M.M. Rahman, and R. Devarajan: Inter. J. Automot. Mech. Eng. Vol. 11(2015), pp. 2272-2281.
[5] M. Mia, M. Al Bashir, A. Khan, N.R. Dhar: Int. J. Adv. Manuf. Technol. Vol. 89 (2017), Issue 1-4, pp. 675–690.
[6] M.K. Gupta, M. Mia, G.R. Singh, D.Y. Pimenov, M. Sarikaya, V.S. Sharma: Int J Adv Manuf Technol. Vol.101 (2019), Issue 1–4, pp 55–69.
[7] M. Mia, M.K. Gupta, J.A. Lozano, D. Carou, D.Y. Pimenov, G. Królczyk G, et al.: J. Clean. Prod. 2019;Vol. 210 (2019), pp.121–33
[8] M.S. Najiha, M.M. Rahman and A.R. Yusoff: Renew. Sustain. Energ. Review, Vol. 60 (2016), pp. 1008-1031.
[9] Y. Songmei, H. Xuebo, Z. Guangyuan, M. Amin: Adv. Prod. Eng. Manage. Vol. 12 (2017), pp. 139-150.
[10] S. Zhang, J.F. Li and Y.W. Wang: J. Cleaner Produc. Vol. 32 (2012), pp. 81-87.
[11] J. Navid: Indian J. Sci. Technol. Vol. 6(2013), pp. 4123-4127.
[12] M.S. Najiha, M.M. Rahman and K. Kadirgama: Inter. J. Automot. Mech. Eng. Vol. 11(2015), pp. 2699-2712.
[13] M.S. Najiha and M.M. Rahman: Inter. J. Automot. Mech. Eng. Vol. 11(2015), pp.2771-2785.
[14] B. Kursuncu, A. Yaras: Int. J. Adv. Manuf. Technol. Vol. 95 (2018), pp. 2005–2013.
[15] M. Kuttolamadom, S. Hamzehlouia, L. Mears: SAE Inter. J. Mater. Manufac. Vol. 3(2010), pp.108-119.

Acknowledgements

The author would like to thanks to Universiti Malaysia Pahang for laboratory facilities and financial assistance under RDU172203 UMP LEAP-3 Flagship project and Postgraduate Research Scheme No. GRS140385.

Strategic Approach to Develop Solutions for Shaping Complex Workpieces of Exotic Materials

Sergio Almeida, John Mo, Cees Bil, Songlin Ding, Xiangzhi Wang

School of Aerospace, Mechanical and Manufacturing Engineering, RMIT University East Campus, Melbourne, VIC 3083, Australia

s3703963@student.rmit.edu.au

Abstract. Exotic materials such as hardened steel and tungsten carbide tool steel have unique resistance and properties that make them hard-to-cut. Thus, research to find better ways to process such materials requires an innovative approach and new ideas. Therefore, the present study investigates the design of a novel WDEM combined with robotic machining to overcome limitations of traditional robotic machining. Wire EDM (WEDM) stands out as a non-traditional machining process able to cut complex profiles of hard-to-cut materials, achieving high dimensional accuracy and superior surface finishing. Unfortunately, WEDM is designed in rigid bed-based CNC machines which restrict design freedom in terms of size, shape and features due to machining envelope constraint. On the other hand, traditional machining processes such as drilling and milling using six-axis industrial robots have been investigated and some applications have successfully delivered cost efficiency, improved envelope and high flexibility. However, due to the structure and strength of the robot arm, accuracy, repeatability and finishing are not comparable to CNC machining outcomes. These researches are also restricted by the power of the robot arm holding the machining tool. This paper explores, identifies and selects suitable configurations and define research actions that must be taken to achieve a highly flexible, accurate machining system for exotic materials.

1. Introduction

The term exotic materials refer to superalloys, ceramics, reinforced plastics, semiconductors, and superconductors. Their application includes cutting tools [1], aerospace, automotive, military applications, medical equipment and bio-implants [2]. However, exotic materials properties are characterised by poor thermal conductivity, high toughness, ultra-hardness, and extremely high work hardening behaviour that combined will place exotic materials under the category of hard-to-cut materials [3].

To machine hard-to-cut materials, conventional processes need to be replaced by electro-discharge machining (EDM). EDM removes electrically conductive material by gradually melting and evaporating portions of the workpiece surface. A series of high-frequency discharges in the dielectric fluid between a conductive workpiece and an electrode by generated heat [4]. In EDM, there are no physical cutting forces between the electrode and the workpiece, avoiding mechanical stresses, chatter and vibrations [5].

EDM machines are numerically controlled. However, CNC machines are characterised by limited working space that often leads the workpiece to be segmented and processed in multiple stages, resulting in deterioration of dimensional precision with a substantial increase in costs and working time [6]. This characteristic makes it difficult to machine complex shapes on EDM. By following EDM principles, a variant process called Wired EDM (WEDM) is found. WEDM cuts the surrounding contours of the workpiece by feeding a wire-electrode that moves as a band saw [7]. To control the wire path, WEDM is configured on computer numerically controlled (CNC) machines. In CNC context, most of the literature has focused on analysing WEDM over different materials at different process conditions [8] usually looking for less surface roughness (SR) or increased material removal rate (MRR) [9].

On the other hand, advantages in 6-axis Industrial Robotic arms (IR) are widely recognised and have been successfully used to replace many manufacturing techniques [10]. IRs with conventional machining tools have many advantages over CNC machines and has attracted significant research interest aiming cost-efficiency, flexibility, design, a multi-functionality [11]. However, IR machining has been severely limited, and most of the current research aims to overcome IR machining problems often originated from lack of IR stiffness and machining vibration [12].

The present research aims to investigate the feasibility of combining EDM with IR to produce complex-shaped workpieces on exotic materials. The research is organised as follows. Section 2 describes the study procedures. Section 3 presents the results and analysis of each field. In section 4, the combination results are presented and discussed; lately, section 5 is the conclusion.

2. Research methodology

To find answers, a systematic review of a two-stage is adopted [13]. Stage 1 search focused on literature in English language from 2009 to 2019. The databases were chosen considering affinity with the subjects of WEDM (1st research axis) and IR machining (2nd research axis), including Springer Link, Science Direct, Scopus, Web of Science, IEEE Xplore, and Google Scholar engine. Stage 2 search focused on industry articles, patents, and reports not include within the academic repositories [14]. To find appropriate keywords, the Web of Science database was first searched, looking for published texts, using the preliminary keywords in Table 1. After reading the titles and abstracts of the ten most cited papers, it was noticed that other strings were better aligned with the research, resulting in a final list of keywords in Table 2.

Table 1. Preliminary research strings

WEDM	IR machining
Exotic material	Machining
Electric discharge	Robotic
Hard to cut	Hard to cut
Wired EDM	

Table 2. Final research strings

WEDM	IR machining
Exotic material	6 axis robot
Hard to cut material	Industrial robot
Electric discharge	Wire cut
Wired EDM	Machining
High-speed WEDM	Grinding

In the 1st research axis (i.e. WEDM), the first selection round resulted in 596 samples. Next, with restriction to each subject, parsing duplicates and categorisation, the sample resulted in 135 contributions. Lately, each paper was reviewed twice. The first review intent to compose the list of criteria that will judge and the second round selects the final literature. Additionally, the references were analysed and papers with potential contributions, not detected in the previous samples, were included in the final review round. The final selected literature on WEDM research axis was shortened to 94 samples. By following the same steps of the 1st research axis, for the 2cd research axis on IR machining, the literature was first screened to 485 papers, next to 108, and finally to 42 samples.

3. WEDM research outcomes and discussion

The literature on WEDM was categorised as a new Method, Process, or a Tool (MPT).

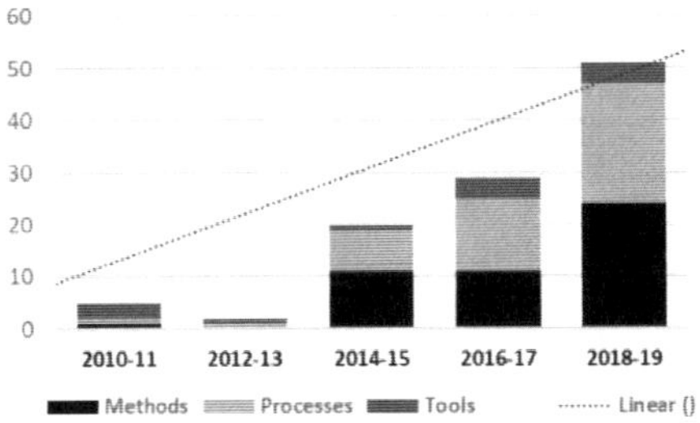

Figure 1. WEDM literature evolution.

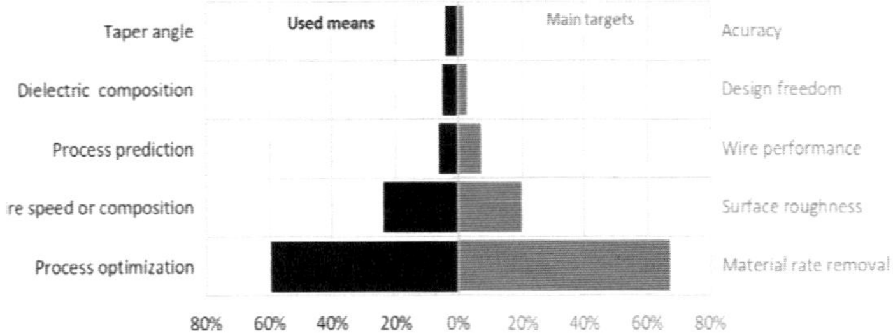

Figure 2. WEDM literature means and targets.

According to Figure 1, new methods and processes optimisation are prominent research fields in WEDM, while new methods are the typical outcome summing 58% of works. However, as in Figure 2, research is mostly process-related, being frequently limited to propose new approaches to optimise or predict WEDM processes parameters based on single or multi-response optimisation [15]. Hence, the derivative research for optimal process parametrisation is the majority and trend of publication, suggesting that new ways to increase process efficiency are necessary. By analysing the use of tools, it was observed that innovative tools are only 9% of research. Moreover, research innovations often focus on wire or dielectric composition, with few exceptions on taper angle and real-time process control, the later, also focusing on process parameters. Therefore, it is possible to suggest that the current literature lacks tools that can improve WEDM application.

In Figure 2, a breakdown of the literature objectives shown that nearly 85% of studies focus on MRR, SR. Regarding simulations, several methods have been identified as able to predict processes with 90 – 95 % accuracy [16] thus, suggesting a high level of maturity with fewer research opportunities. Except for Micro-WEDM [17] and taper angle issues [18], no research was found focusing on dimensional accuracy. Therefore, is possible to suggest that conventional WEDM is delivering sufficient accuracy for straight cutting; however, the design is still limited by CNC constraints, and current solutions deliver reduced shape and accuracy [18, 19].

Regarding wire advances, higher performance wires rely on complex shaping and rare metals resulting in high-cost, damage to the wire chopper, straightness issues [20]. On the other hand, the evolution of pure brass wire to core brass-alloys next added with reduced Zinc content in coating alloys provides enhanced cooling and flushability and core brass wires persist as the best trade-off for cost, cut speed, surface roughness, and endurance [21]. As a solution to the high wire cost, an increasing interest in high-speed WEDM (HS-WEDM) is noticed. However, in HS-WEDM, when the MRR is near to 150 mm^2/min, the workpiece surface presents burns and frequent wire breakage [22].

The common ways to improving MRR and SR are cooled fluids [23] or additives [24], also flushing combined with ultrasonic field [4]. To solve surface burning, a new dielectric fluid with higher vaporisation point was developed for HS-WEDM with increased MRR of 330 mm^2/min, under higher average current of 15A and stable machining [22]. Concerning to ultrasonic combination, the principle behind the wire electrode activated with ultrasonic is to reduce or even eliminate the undesired effects of the electromagnetic field, resulting in higher dimensional and shape accuracy [25]. Regarding flush operation, the ultrasonic field helps to cope with the effects of gravity when the wire is working on non-vertical straight cutting [26], therefore, useful to improve design freedom, higher MRR and improved SR [25].

Lately, all found patents refer to wire development while only one journal article is directly related to the industry where a CNC machine is improved, and advantages argued on taper angle, increased workpiece thickness, and automatic wiring [19]. This distribution, combined with Figure 2 suggests that research focusing on new tools rather than processes parameters may be more relevant to the current industry needs.

4. IR research outcomes and discussion

Regarding IR machining, the criteria adopted to select the final literature sample were new MPTs.

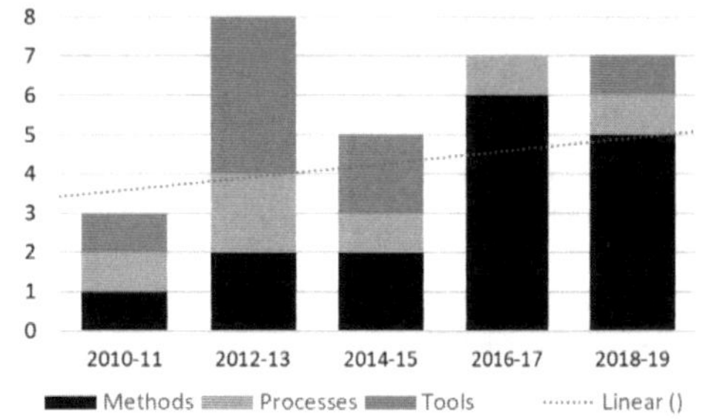

Figure 3. IR machining literature evolution.

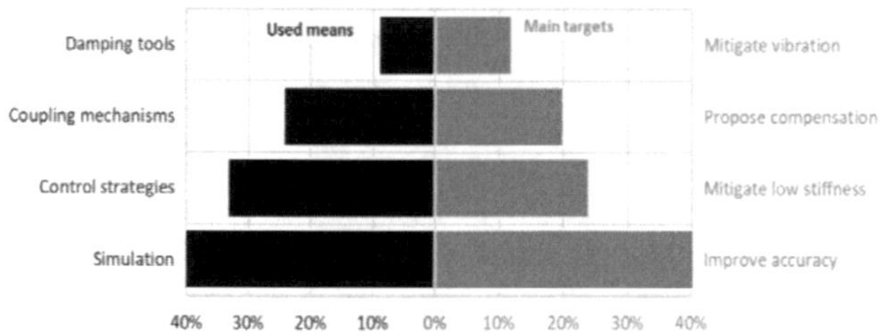

Figure 4. IR literature means and targets.

As in Figure 3, IR machining is a growing research field, while new methods are the most frequent topic (50%), followed by new tools (31%) and new processes (19%). Besides, it was found that nearly all publications keep aiming to improve stiffness and suppress machining vibration [11]. However, as in Figure 4, solving the lack of stiffness and machining vibration are ways to improve IR machining accuracy. Since substantial research exist aiming to compensate errors, it is implied that the lack of stiffness and machining vibrations are problems not solved. Meanwhile, few studies were found focusing on machining efficiency in terms of MRR or SR [6]. On the contrary, improvements of MRR and SR are secondary effects observed when vibration, stiffness and simulation are managed [27], thus, suggesting that new ways to solve IR limitations must be achieved upon efficiency.
As in Figure 4, the main approaches to improve IR machining can be found in four groups. The first and more common relies on simulations that allow offline programming [28] to convert 3D workpiece into cutting program [29] and avoid trajectory collision as well as IR singularities [30]. Hence, simulations deliver robot pose for maximised stiffness and embedded error compensation tailored for IR model and manufacturer [31].

5. Combination results and discussion

To identify feasibility, advantages and limitations, for both, WEDM and IR, the strengths and weakness were extracted and combined in Table 3, resulting in four scenario-quadrants.

Table 3. Combination matrix of IR & WEDM

		Wire EDM in CNC machines		
		Strengths +WE1. High accuracy +WE2. High SR quality +WE3. Hard to cut material +WE4. No vibration or forces	*Weakness* -WE1. Low MRR -WE2. Low design freedom -WE3. Limited envelope -WE4. Expensive wire usage	(+) Desirable
Machining Robots	*Strengths* +IR1. Large envelope +IR2. Design freedom +IR3. Path programming +IR4. Easier sensing	Strengths combination: IR machining of hard-to-cut materials, higher accuracy, larger envelop, easier path programming, sensing processes, no vibration and stiffness problems.	IR strengths & WEDM weakness: Same MRR, increased design freedom, flexible configuration, easier sensing, and similar wire cost.	
	Weakness -IR1. Low stiffness -IR2. Low precision -IR3. Easy to cut materials -IR4. Limited EE weight	WEDM strengths & IR weakness: WEDM accuracy to be confirmed, while end-of-effect WEDM tool will demand a new and lighter design than those usually found in CNC machines.	Weakness combination: It is prone to lower WEDM accuracy, yet with not improved wire cost process.	Undesirable (-)
		(+) Desirable →	Undesirable (-)	

Due to the combined strengths, the resulting process should achieve improved design freedom, enlarged machining envelops, and flexible processes configurations potentially assisted by IR cut programming as well as sensing applied for WEDM parameters. Another two quadrants on weakness and strengths reveal some risks and many symbioses. One could argue in certain WEDM dimensional degradation; however, this conclusion is not straightforward. As most of IR dimensional error is originated by conventional machining forces and vibration, the WEDM characteristic of non-contact and nearly zero forces shall improve IR precision machining into a level yet to be confirmed, since a few papers have quantified machining error sources coming from the IR exclusively. For WEDM, kinematics is recognised as a critical process parameter; thus, adapting WEDM into an IR shall enable electrode movement with superior smoothness and SR. As a risk, the WEDM feeding system is complex and potentially heavy to act as an EE. In this sense, a possible solution is to adopt HS-WEDM type that reuses the wire and thus, presents more possibilities to be reconfigured in a lighter design and yet deliver reduced costs for WEDM. Lately, since many weaknesses on WEDM are amended by IR and vice versa, most combinations do not necessarily result in deteriorated outcomes. For instance, low MRR of WEDM is prone not to be improved by IR; however, the low MRR will demand to the IR a low level of speed, which is desirable since high speed is a source of IR dimensional error.

6. Conclusion

By analysing the strengths and limitations of IR machining and WEDM, a combination was investigated, and possible outputs were identified. One of the key findings is that no previous studies have focused on combining WEDM into an IR. Lately, the results confirm that the combination is potentially feasible, and several problems found in IR machining are expected to be dramatically reduced by WEDM combination. Future research to this work is to design a new WEDM end effector, by leverage and adapt existing methods and tools associated with IR machining and WEDM to develop a hybrid approach that aligns those two fields with a specific focus on hard-to-cut metals.

7. References

[1] Shen Y, Liu Y H, Dong H, Zhang K, Lv L, Zhang X Z, Wu X L, Zheng C and Ji R J 2017 Surface integrity of Inconel 718 in high-speed electrical discharge machining milling using air dielectric *International Journal of Advanced Manufacturing Technology* **90** 691-8

[2] KumarNishasoms K R 2019 Desirability-Based Multi-objective Optimization and Analysis of WEDM Characteristics of Aluminium (6082)/Tungsten Carbide Composites *Arabian Journal for Science and Engineering* **44** 893-909

[3] Rahul, Abhishek K, Datta S, Biswal B B and Mahapatra S S 2017 Machining performance optimization for electro-discharge machining of Inconel 601, 625, 718 and 825: an integrated optimization route combining satisfaction function, fuzzy inference system and Taguchi approach *Journal of the Brazilian Society of Mechanical Sciences and Engineering* **39** 3499-527

[4] Nani V-M 2017 Complex phenomena study in dielectric fluid from gap during the W-EDM processing in ultrasonic field *International Journal of Advanced Manufacturing Technology* **92** 197-215

[5] Czelusniak T, Higa C F, Torres R D, Laurindo C A H, de Paiva Júnior J M F, Lohrengel A and Amorim F L 2018 Materials used for sinking EDM electrodes: a review *Journal of the Brazilian Society of Mechanical Sciences and Engineering* **41** 14

[6] Mohammad A E K and Wang D 2015 A novel mechatronics design of an electrochemical mechanical end-effector for robotic-based surface polishing. In: *System Integration (SII), 2015 IEEE/SICE International Symposium on*: IEEE) pp 127-33

[7] Dixit M D S 2018 Advanced Machining Processes

[8] Yilmaz O, Bozdana A T and Okka M A 2014 An intelligent and automated system for electrical discharge drilling of aerospace alloys: Inconel 718 and Ti-6Al-4V *The International Journal of Advanced Manufacturing Technology* **74** 1323-36

[9] Rahim M Z, Li G, Ding S, Mo J and Brandt M 2015 Electrical discharge grinding versus abrasive grinding in polycrystalline diamond machining-tool quality and performance analysis *International Journal of Advanced Manufacturing Technology* **81** 1-4

[10] Keating S and Oxman N 2013 Compound fabrication: A multi-functional robotic platform for digital design and fabrication *Robotics and Computer-Integrated Manufacturing* **29** 439-48

[11] Ji W and Wang L H 2019 Industrial robotic machining: a review *International Journal of Advanced Manufacturing Technology* **103** 1239-55

[12] Schneider U, Ansaloni M, Drust M, Leali F and Verl A 2013 Experimental Investigation of Sources of Error in Robot Machining. (Berlin, Heidelberg: Springer Berlin Heidelberg) pp 14-26
[13] Verhagen W J C, Bermell-Garcia P, van Dijk R E C and Curran R 2012 A critical review of Knowledge-Based Engineering: An identification of research challenges *Advanced Engineering Informatics* **26** 5-15
[14] Ely C and Scott I 2007 *Essential study skills for nursing*: Elsevier Health Sciences)
[15] Kamei T, Okada A and Okamoto Y 2016 High-speed Observation of Thin Wire Movement in Fine Wire EDM *Procedia CIRP* **42** 596-600
[16] Devarasiddappa D, George J, Chandrasekaran M and Teyi N 2016 Application of Artificial Intelligence Approach in Modeling Surface Quality of Aerospace Alloys in WEDM Process *1st Global Colloquium on Recent Advancements and Effectual Researches in Engineering, Science and Technology - Raerest 2016* **25** 1199-208
[17] Chavoshi S Z, Goel S and Morantz P 2017 Current trends and future of sequential micro-machining processes on a single machine tool *Materials & Design* **127** 37-53
[18] He H Y L L L 2017 Large taper mechanism of HS-WEDM *International Journal of Advanced Manufacturing Technology* **90** 2969-77
[19] Takayama Y, Makino Y, Niu Y and Uchida H 2016 The Latest Technology of Wire-cut EDM *Procedia CIRP* **42** 623-6
[20] Kapoor J, Singh S and Khamba J S J P o t I o M E, Part B: Journal of Engineering Manufacture 2012 High-performance wire electrodes for wire electrical-discharge machining–a review **226** 1757-73
[21] Ciwen H, Jinsheng Z and Jianyong L 2018 Ultra-long Wire Reciprocated-WEDM with Dual Tensile Reels Winded *Procedia CIRP* **68** 115-9
[22] Pan H W, Liu Z D, Li C R, Zhang Y Q and Qiu M B 2017 Enhanced debris expelling in high-speed wire electrical discharge machining *International Journal of Advanced Manufacturing Technology* **93** 2913-20
[23] Gov K 2017 The effects of the dielectric liquid temperature on the hole geometries drilled by electro erosion *International Journal of Advanced Manufacturing Technology* **92** 1255-62
[24] Kumar S, Mitra B, Dhanabalan D S J I J o M, IT and Engineering 2018 The state of art: revolutionary 5-axis CNC wire EDM and its recent developments **8**
[25] Nani V-M 2017 The ultrasound effect on technological parameters for increase in performances of W-EDM machines *International Journal of Advanced Manufacturing Technology* **88** 519-28
[26] Mouralova K, Klakurkova L, Matousek R, Prokes T, Hrdy R and Kana V 2018 Influence of the cut direction through the semi-finished product on the occurrence of cracks for X210Cr12 steel using WEDM *Archives of Civil and Mechanical Engineering* **18** 1318-31
[27] Chen F, Zhao H, Li D W, Chen L, Tan C and Ding H 2019 Robotic grinding of a blisk with two degrees of freedom contact force control *International Journal of Advanced Manufacturing Technology* **101** 461-74
[28] Leali F, Pellicciari M, Pini F, Berselli G and Vergnano A 2013 An Offline Programming Method for the Robotic Deburring of Aerospace Components. (Berlin, Heidelberg: Springer Berlin Heidelberg) pp 1-13
[29] Søndergaard A, Feringa J, Stan F and Maier D 2018 Robotic abrasive wire cutting of polymerized styrene formwork systems for cost-effective realization of topology-optimized concrete structures *Construction Robotics* **2** 81-92
[30] Mao Y, Zhao H, Zhao X and Ding H 2017 Trajectory and Force Generation with Multi-constraints for Robotic Belt Grinding. (Cham: Springer International Publishing) pp 14-23
[31] Xie H, Li W and Yin Z 2018 Posture Optimization Based on Both Joint Parameter Error and Stiffness for Robotic Milling. (Cham: Springer International Publishing) pp 277-86

8. Acknowledgements

The project is supported by Natural Science Foundation of Jiangsu Province, China (Grant No. BK20150255)

IOP Conf. Series: Materials Science and Engineering **831** (2020) 012004 doi:10.1088/1757-899X/831/1/012004

Case study: Optimization of Case Depth in Induction- hardened 42CrMo Steel Shaft

R SYED [1, 2], J G WANG [1], D Y Jing [2], X L Liu[2], Y M Yang[2], M F Panhwar[3], K Wang[2] and Q Y Chen[2]

[1] School of Mechanical Engineering, Zhejiang University Yuquan Campus, #38 Zheda Road Hangzhou, P R China 310027

[2] Zhejiang Shuanghuan Driveline Co., Ltd. Yintao Road#201, Bingang Industrial Zone, Shamen Town, Yuhuan City, Taizhou city, Zhejiang, P R China 317600,

[3] Central China Normal University, Wuhan, Hebei, P R China

syedrizwanulhaq@hotmail.com / rizwan@gearsnet.com

Abstract. Low alloyed high-grade chrome-molybdenum ferritic steel transmission output shaft was investigated. The medium-carbon steel which is commonly used induction hardened transmission components and drive shafts in the automotive industry. The aim of this experiment is achieve the case depth that spline shaft design requirement. The required case depth for design the proper induction hardening coil and varying the parameter of it has been achieved. The results agree well with literatures, that optimum case depth maximizing the strength and fatigue life of spline shaft.

1. Introduction

Heavy machinery splined shafts commonly used to transfer the high torque, choice of hardening process is crucial to achieve an optimal strength of the part[1]. Induction hardening heat treatment commonly applies for spline shafts, which enhance the torque capacity of the shaft. The strength of the spline connection mainly depends upon material strength and spline design. Last decade several researches published have defined spline shaft and non-spline shaft's strength, hardness depth, fatigue life and failure. These parameters may be able to optimize by adjusting the parameters of induction hardened process parameter [2]–[8].

The process of induction hardening has been carried out for long time by several researchers. However, its application for studying the hardness profile sensitivity on the material or coil design and/or as a function of machine parameters are still rather rare. Furthermore, the effect of machine parameters and coil design on the hardness profile has limited studied and documented[2].

The analysis presented in this case study to optimize the hardness depth from hardening process of heavy duty machinery splined shaft by taking into account the detailed effect of the geometry of spline.

2. Geometry of the splined shaft

In this case study uses the transmission splined shaft 260 mm in length with both ends splined. Splined shaft parameters are tabulated in table 1. Figure.1 pictorial cross-section transmission splined shaft used in this case study.

Table 1. Splined Data

Spline	No. of Teeth	Pressure angle	Major diameter (mm)	Minor diameter (mm)
A	22	30°	72.28-72.25	65.633
B	17	30°	63.73-63.7	56.134

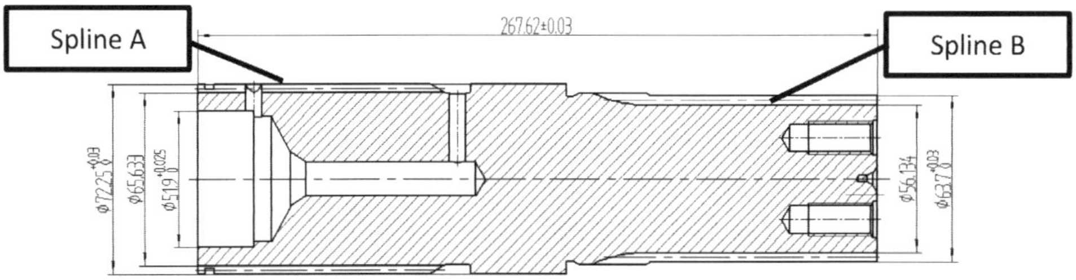

Figure 1. Splined Shaft Cross-section View

3. Mechanical Properties of 42CrMoH

The 42CrMoH (SAE 4140) medium-carbon steel, it is commonly used in automotive industry. This material in suitable for induction hardened components [8], [9]. The chemical composition and mechanical properties tabulated in Table 2.

Table 2. Chemical Composition and Mechanical Properties of 42CrMoH (SAE 4140)

Chemical composition									
Element	**C**	**Si**	**Mn**	**P**	**S**	**Cr**	**Mo**	**Ni**	**Ti**
Value	0.449	0.26	0.79	0.019	0.003	1.17	0.19	0.03	0.0019
Hardenability									
Technical Specification		**J5**		**J9**				**J15**	
Value		56.99		55.42				51.25	

4. Experiment

4.1. Sample Preparation and Hardness profile

In this case study, medium carbon steel rods hardening & tempering to achieve the hardness (HRC 28.5, 28.7, 27.9, 28.9). These four samples were machined as per the print requirement. Samples simplified CAD model as shown in Figure 1, and detail of the spline tabulated in table 1.

IOP Conf. Series: Materials Science and Engineering **831** (2020) 012004 doi:10.1088/1757-899X/831/1/012004

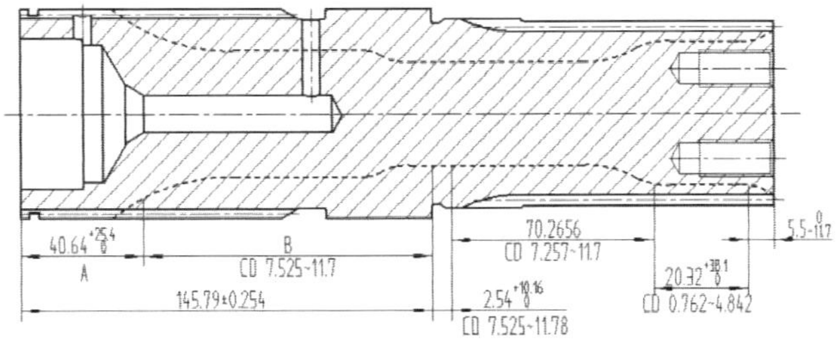

Figure 2. Hardness profile requirement

Furthermore, the hardness profile requirement defined in the Figure 2 and required case depth tabulated in table 3 and finished part hardness requirement ≥ 48HRC.

Table 3. Case Depth Requirement

Section	Speciation	Value
A	Width	40.64-66.04mm
B	Case Depth (CD)	7.525-11.7mm
C	Case Depth (CD)	7.525-11.78mm
D	Case Depth (CD)	7.257-11.7mm
E	Case Depth (CD)	0.762-4.842mm
F	Width	0-11.7mm

4.2. Design and Process Considerations

It is critical not only to meet the hardened depth requirement, but also prevent excessive heating in such area as, B, C, D, and E. Excessive heat in these regions may cause the crack on the surface and lead to excessive distortion.

The inductor needs to be designed carefully to prevent cracking possibility and excessive distortion issues. This part has a 0.762- 4.842mm min pattern depth. Typically, it would use 10 KHz for anything up to 6.35 mm case depth. Heating beyond that depth loses the benefit of frequency penetration and requires heating by conduction. These results in slower travel that can lead to surface overheat. For effective heating at this depth would typically choose a machine with a nominal output frequency of 3 KHz. That frequency will effectively penetrate to the minimum depths required resulting in faster and minimizing the chance for surface overheat. With these deeper case depths quench capacity is also a concern. Due to the surface area parts like this are typically scanned due to power constraints with a machined Integral quench (MIQ) coil Figure 3.

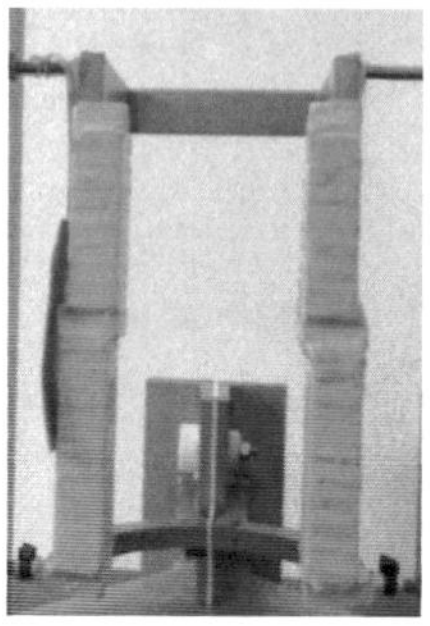

Figure 3. Induction coil design

5. Results and Discussion

The results from the experiment of the two samples are measure with venire caliper. These two samples of spline shaft result compare the case depth, which defined in Figure 4 and data tabulate in Table 4.

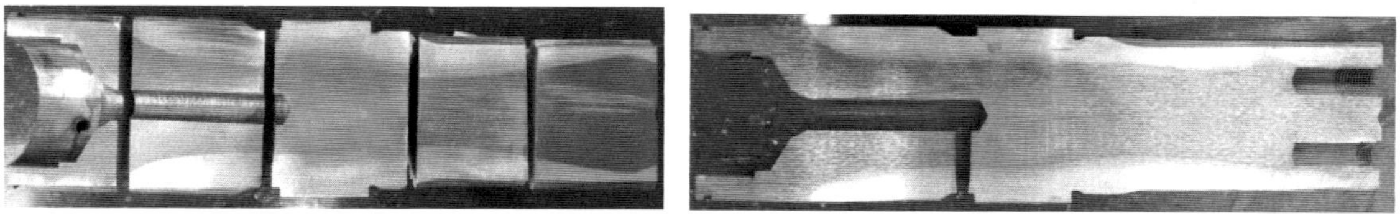

Figure 4. Hardness Profile and Depth, left side sample#1, and right side is Sample #2.

Table 4. Case depth Comparison

Section	Specification	Value	Sample #1	Sample#2
A	Width	40.64-66.04mm	43.9 mm	43.75 mm
B	Case Depth (CD)	7.525-11.7mm	12.1-13.5 mm	9.05-9.09 mm
C	Case Depth (CD)	7.525-11.78mm	12.5 mm	8.6 mm
D	Case Depth (CD)	7.257-11.7mm	11.9 mm	9.2 mm
E	Case Depth (CD)	0.762-4.842mm	4.34-5.6 mm	2.73- 4.34 mm
F	Width	0-11.7mm	0 mm	0 mm

The induction heat treatments were conducted to achieve required case depth by varying the frequency 4 KHz to 8 KHz and distance between coil and small diameter. The hardened surface layers (cases) of the undercut, spline and flat segment of these samples are presented in Figure 4. The non- transformed core and the transition layer between the case and the core can also distinguish. The higher case depths achieved by increased heating power.

Figure 5. Sample # 2 Microstructure at 500 x, at left show surface, at right shows core

In detail, the case layers of sample#2 (as it fulfills the requirement) have martensitic microstructure (Figure 5). The transition layers are characterized by microstructures composed of martensite and ferrite.

6. Conclusion

The current case study presents an induction hardening spline shafts by taking into account the case depth of spline shaft. It can be concluded that required case depth at vary area (undercut, under spline teeth, varying diameter and transition of diameter) achieved by 3 KHz to 8 KHz with help of special designed induction coil. The nominal required case depth value without surface crack has been achieved. The optimum hardness depth for hardened shaft obtained nearly to the half of the shaft radius. As the subsurface crack initiation with increase the hardness depth due to increase of residual stresses in the core, which is detrimental for fatigue life.

7. Reference

[1] I. Barsoum, F. Khan, and Z. Barsoum, "Analysis of the torsional strength of hardened splined shafts," *Materials and Design*, vol. 54, pp. 130–136, 2014.

[2] N. Barka, "Study of the machine parameters effects on the case depths of 4340 spur gear heated by induction — 2D model," 2017.

[3] H. Hammi, A. E. Ouafi, and N. Barka, "Study of Frequency Effects on Hardness Profile of Spline Shaft Heat-Treated by Induction," no. March, pp. 1–9, 2016.

[4] A. Lasaosa, K. Gurruchaga, F. Arizti, and A. Martinez-de-guerenu, "Induction Hardened Layer Characterization and Grinding Burn Detection by Magnetic Barkhausen Noise Analysis," *Journal of Nondestructive Evaluation*, pp. 1–7, 2017.

[5] R. Goldstein, J. Jackowski, G. Fett, D. C. Technology, A. Hills, and D. Corporation, "MODELING STRESS AND DISTORTION OF FULL-FLOAT TRUCK," no. 1, pp. 1–9.

[6] C. He, M. Yang, X. Liu, X. Wang, and B. Wu, "Characterisation of case depth in induction-hardened medium carbon steels based on magnetic minor hysteresis loop measurement technique," *Philosophical Magazine*, vol. 6435, no. August, pp. 1–16, 2017.

[7] P. D. H. Burande, "A review: Induction Hardening on Axle shaft," vol. 35, no. 1, pp. 43–46, 2016.

[8] C. Ondrusek, "Magnetic properties of 42CrMo4 steel," 2017.

[9] Y. Li, S. Zhao, S. Fan, and G. Yan, "Study on the material characteristic and process parameters of the open-die warm extrusion process of spline shaft with 42CrMo steel," *Journal of Alloys and Compounds*, vol. 571, pp. 12–20, 2013.

[10] G. A. Fett and D. Corp, "IMPORTANCE OF INDUCTION HARDENED CASE DEPTH IN," 2009, no. October, pp. 15–19.

[11] A. Tjernberg, "Fatigue lives for induction hardened shafts with subsurface crack initiation," *Engineering Failure Analysis*, vol. 9, no. 1, pp. 45–61, 2002.

[12] S. B. Leen, I. R. McColl, C. H. H. Ratsimba, and E. J. Williams, "Fatigue life prediction for a barrelled spline coupling under torque overload," *Proceedings of the Institution of Mechanical Engineers, Part G: Journal of Aerospace Engineering*, vol. 217, no. 3, pp. 123–142, 2003.

Acknowledgments

This work supported by Research Program supported by the Science & Technology Bureau of Yuhuan, and the Taizhou city government, Zhejiang P.R. China.

IOP Conf. Series: Materials Science and Engineering **831** (2020) 012005 doi:10.1088/1757-899X/831/1/012005

Low Frequency Vibration Isolation Performance of the Plate with Periodic Cylindrical Oscillators

Peng Zhou, Shui Wan *

School of Transportation, Southeast University, Nanjing 211189, China

zhoupeng_phd@163.com

Abstract: The paper is devoted to the vibration reduction and isolation performance of a periodic structure with cylindrical vibrators. A plate with periodic oscillators was investigated using finite element technic. In the finite element analysis, test specimens were modelled with finite element software. With appropriate geometric and material parameters, the structure can obtain elastic wave band gaps, where vibrations are prohibited. So, the structure can be regarded as a kind of vibration isolation material. The dispersion curve and transmission spectrum of the structure were calculated to find the range of the band gaps. Both the longitudinal vibration and flexural vibration of the plate are studied. The results show that the designed plate with proper parameters has a low frequency range of band gaps. It has a wide application prospect in the field of low frequency vibration isolation.

1. Introduction

During the past decade, propagation of elastic wave in periodic structures, so-called Phononic crystals (PCs), has received considerable attention for their potential applications in vibration isolation [1][2][3]. The structure has elastic wave band gap, in which the elastic wave propagation is prohibited. If we change the geometric or material parameters of the periodic structure, the position of the band gap will also change. Over the past decades, a significant amount of researches have been studied on the band gap characteristics of periodic structure. Periodic structure can be divided into the types of Bragg scattering and local resonance according to the band gap formation mechanism.

The periodic structure based on Bragg scattering mechanism is studied firstly. Vasseur et al.[4] tested the transmission spectrum of two-dimensional solid/solid triangle lattice phononic crystal in the direction of ΓX and ΓJ. The plane wave expansion method and the theory of finite difference time domain method is used for validation. The wavelength of band gap frequency of Bragg scattering type corresponds to the same order of magnitude as lattice constant. The first locally resonant phononic crystal proposed by professor Liu et al.[5] It is pointed out that the wave length corresponding to the low frequency band gap is larger than its lattice constant through a three-component periodic structure (epoxy resin-soft rubber-lead), that is, the low-frequency band gap can be obtained with a smaller structure size.

In recent years, more and more researches focus on vibration suppression of periodic structural plates. Pennec et al [6]. found that the opening of its gap requires appropriate geometrical parameters by the investigation on a two-component pillared PC plate. Hsu [7] proposed the pillars with a "neck" in the PC slab with stepped resonators and proved that there are both local resonance band gaps and Bragg band gaps in the pillars. Zhao et al.[8][9] studied the vibration band gaps in pillared phononic crystal plate and a series of parameter analysis was carried out. However, at present, phononic crystal plates are

generally small in size and high in vibration isolation frequency, and are mainly used in aviation and mechanical fields.
In this paper, a large scale periodic structural plate with a low frequency band gap is proposed. Therefore, it can be more applicable to other low frequency vibration isolation fields. The longitudinal and flexural vibration band gaps of the periodic structure plate is concluded throughout the calculation of dispersion curve of cell structure and analysis of its displacement mode. The longitudinal and flexural transmission spectrum is also calculated for verification.

2. Model and method

2.1 Model

Figure 1 depicts the plate with periodic vibrator pillars. The oscillator is arranged in 6 rows and 6 columns periodically on the plate. The unit cell of the periodic structure is shown in Figure 2. The vibrator pillar in the unit cell is composed of two layers of cylindrical slices of rubber and steel. The diameter of the rubber layer and steel layer are d_1 and d_2 respectively. Since the rubber layer is generally seen like a spring, d_1 is set smaller than d_2 to save materials. Treating one rubber stub and one steel cap as a vibrator, such vibrators are periodically arranged on the surface of the plate. The geometric parameters of the structure are defined as follows: The distance between adjacent cells (lattice constant) is a, so the plate is $6a$ in length and $6a$ in width. The thickness of the plate is denoted by e; and the thickness of the rubber layer and steel layer are h_1 and h_2 respectively. The z-axis is perpendicular to the plate. All materials used in this study are assumed to be elastically isotropic for simplicity. The materials parameters used in the calculation are shown in Table 1.

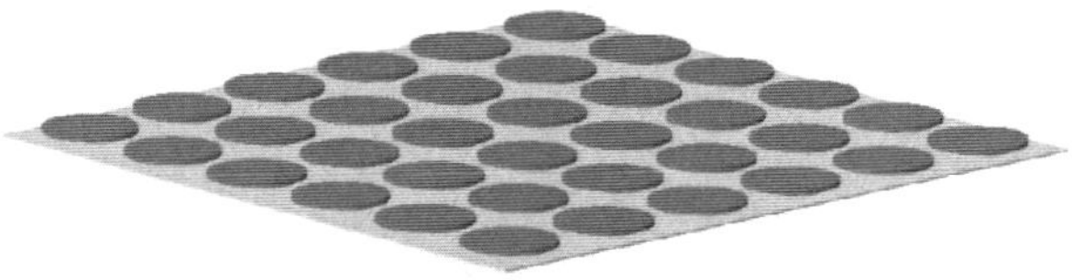

Figure 1. The plate with periodic cylindrical oscillators

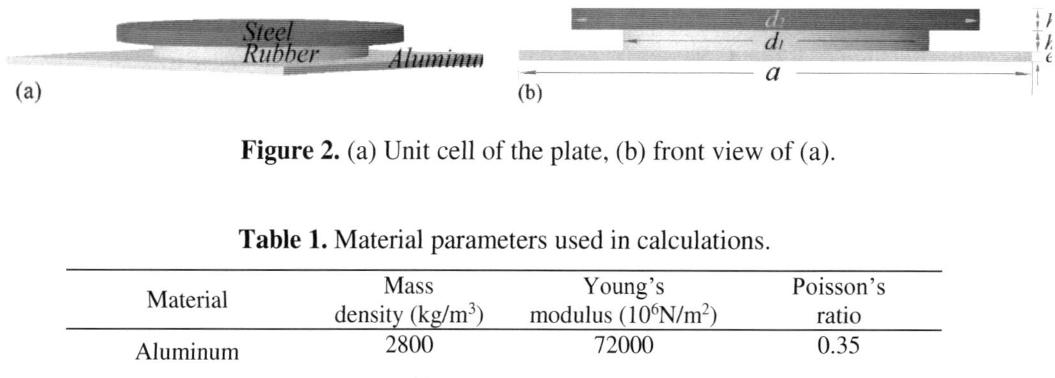

Figure 2. (a) Unit cell of the plate, (b) front view of (a).

Table 1. Material parameters used in calculations.

Material	Mass density (kg/m^3)	Young's modulus (10^6N/m^2)	Poisson's ratio
Aluminum	2800	72000	0.35
Steel	7780	210000	0.3
Rubber	1300	0.118	0.47

2.2 Method

In order to obtain the dispersion curve and frequency band gap of the plate, the finite element method is used to calculate the characteristic frequency and resonance mode, which has been proven to be an efficient method in previous research [8]. Compared with other analytical methods, FEM is easier to solve PCs with complex shape of scatters. Considering the periodicity of the structure, only one cell is needed to calculate the dispersion curve. In the finite element algorithm, the discrete form of the eigenvalue equations in the unit cell can be written as

$$(\mathbf{K}\text{-}w^2\,\mathbf{M})\,\mathbf{u} = 0 \tag{1}$$

where **K** and **M** are the stiffness and mass matrices of the unit cell respectively, **u** is the displacement of the nodes and w is the circular frequency. Computations are performed with software COMSOL Multiphysics. The unit cell is meshed with the "fine" grade of the predefined mash sizes to get good convergence. Free boundary conditions are applied to the top and bottom surfaces of the unit cell. With the Bloch-Floquet theorem, periodic boundary conditions are used for the interfaces between the nearest unit cells:

$$u_j(x+a, y+a) = u_j(x,y)e^{\mathrm{i}(k_x a + k_y a)}, (\mathrm{j} = \mathrm{x}, \mathrm{y}, \mathrm{z}) \tag{2}$$

where u_j is the elastic displacement vector; (x, y, and z) are the position vectors; k_x and k_y are Bloch wave vectors limited in the irreducible first Brillouin zone (BZ). The dispersion relations can be obtained by changing the value of k_x and k_y along the boundary of the first Brillouin zone and calculating the eigenvalue problem generated by the FEM algorithm.

3. Numerical results and analyses

During the calculation, lattice constant a =1m, plate thickness e =0.02m, h_1=h_2=0.04m and d_1=0.6m, d_2=0.8m were kept unchanged. The dispersion curve of the unit cell is reported in Figure 3. The propagation of multiple vibrations is coupled in this dispersion curve. We need to find out the starting frequency fs (from which frequency the gap starts) and the cutoff frequency fc (at which frequency the gap ends) of the flexural vibration band gaps and the longitudinal vibration band gaps. We can pick them out by distinguish the modes from the others. The displacement fields of the eigenmodes labelled in Figure 3 are shown in Figure 4.

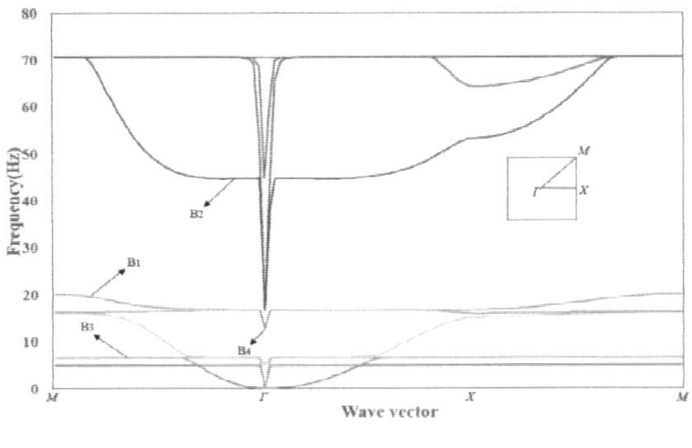

Figure 3. Dispersion curve of the unit cell

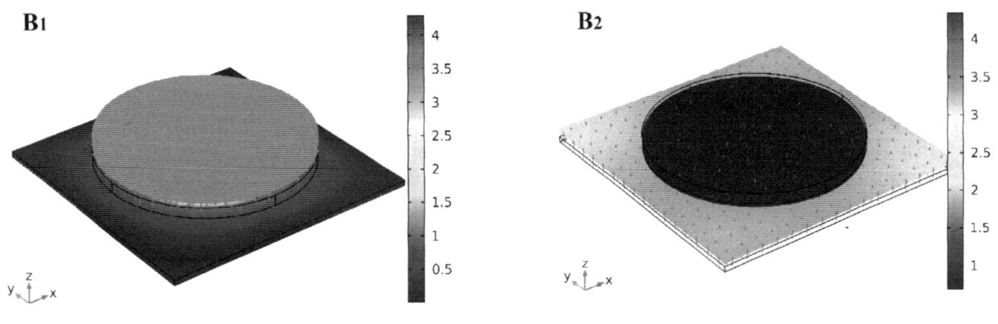

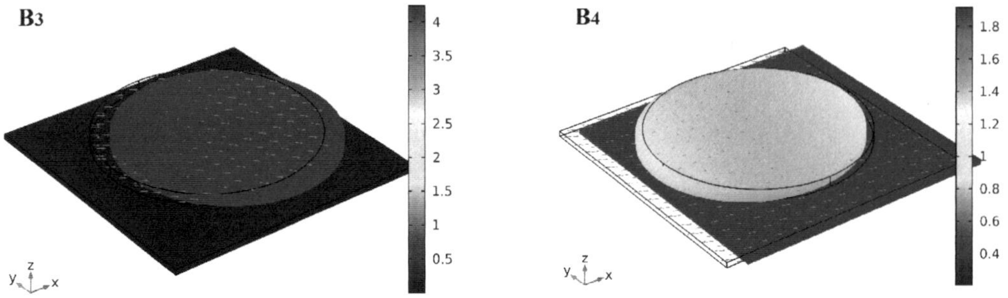

Figure 4. Displacement field of eigenmode at point B_1, B_2, B_3 and B_4.

Mode B_1 (18 Hz) concentrates the vibration energy in the steel cap in the vibrator pillar which vibrates along the z-axis. This indicates that flexural vibration band gaps begin to form. So, this mode corresponds with the starting frequency of the flexural vibration band gap (FVBG). In mode B_2 (44 Hz), the vibration of the base plate and the steel cap is inverse along the z-axis to achieve the dynamic balance. This mode corresponds with the cut-off frequency of the FVBG. The dominant vibration of B_1 and B_2 are along the z-axis and the vibration couples with the antisymmetric Lamb mode of the base plate, as a result the modes are involved in the formation of the FVBGs.

For longitudinal vibration band gap (LVBG), we follow the same analytical thinking. In mode B_3 (7 Hz), the vibration energy concentrates on the whole body movement of the vibrator pillar along the xy plane and keep their base plates stationary, which indicates the start of a LVBG. While Mode B_4 (13 Hz) achieves dynamic balance of the inverse vibration of the lateral steel cap and the base plate in the xy plane. So, Modes B_3 and B_4 correspond with the starting and cut-off frequencies of the LVBG. The dominant vibration of B_3 and B_4 are in the xy plane and the vibration couples with the symmetric Lamb mode of the base plate, as a result the modes are involved in the formation of the LVBGs.

Through the analysis of the above band structure mode, we found that the FVBGs start from mode B_1 at 18 Hz and cut-off at mode B_2 at 44 Hz. The LVBGs start from mode B_3 at 7 Hz and cut-off at mode B_4 at 13 Hz similarly.

To further demonstrate the existence of the FVBG and LVBG of the studied structure, we also calculated the transmission power spectrums. Firstly, a displacement excitation d_{in} was imposed on the middle of one end of the plate, and set up the domain point probe on the middle of the opposite end of the plate (Figure 5), the displacement response d_{out} was picked up. The transmission spectra L_{d} is defined as

$$L_{\mathrm{d}} = 20\log\frac{|d_{\mathrm{out}}|}{|d_{\mathrm{in}}|}. \tag{3}$$

Where d_{in} and d_{out} are the vibration displacement of the input and output sides of the plate, respectively. The response spectra of the plate are shown in Figure 6.

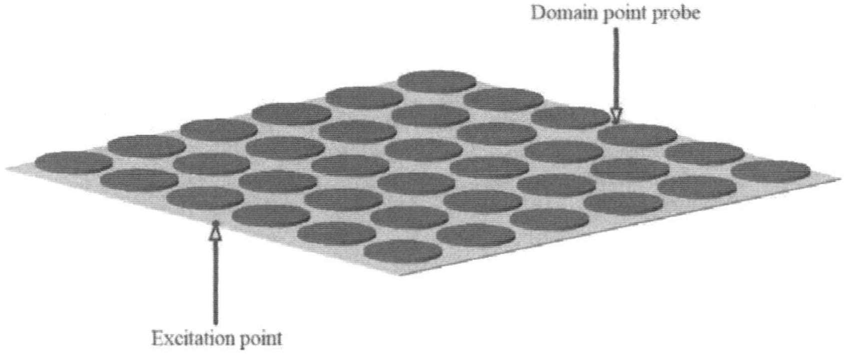

Figure 5. Layout of measuring points of calculated transmission power spectrum

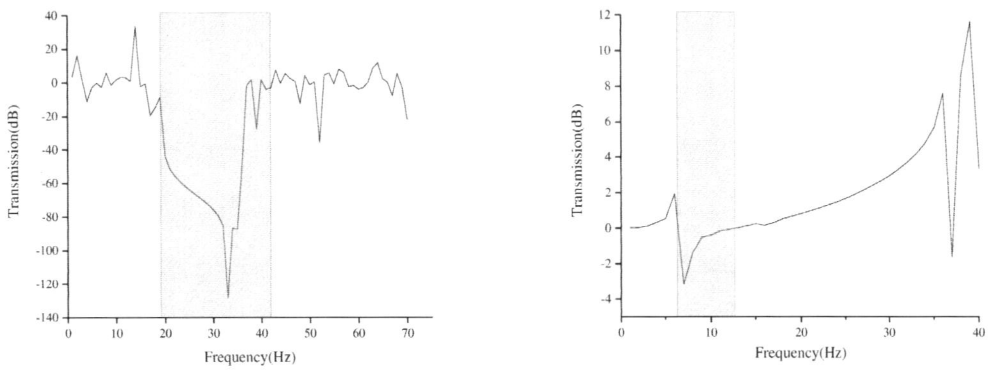

Figure 6. Transmission power spectrum of flexural vibration and longitudinal vibration propagating in the plate. The gray shadow regions denote the band gaps.

The transmission spectrum depicts that the band gap range of flexural vibration is between 19 Hz to 43 Hz. Meanwhile, the longitudinal vibration enjoys the band gap between 7 Hz to 13 Hz. The frequency range of the attenuated transmission spectrums of the vibrations is nearly consistent with the results obtained from the band diagram (Figure 3). The correctness of the calculation results is further proved. In order to show the propagation of flexural vibration with different frequencies in the plate, the displacement modes of the plate within the band gap and out of the band gap are plotted (Figure 7). It can be clearly seen that the vibration can pass through the plate without any loss if its frequency is outside of the band gap. However, the vibration within the band gap will be significantly suppressed when it propagates in the plate, and the vibration displacement is mainly limited near the excitation point.

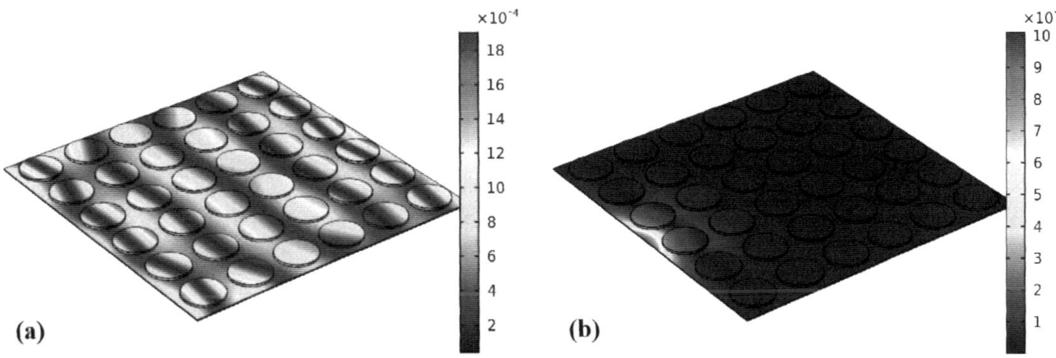

Figure 7. The modes corresponding to the frequencies of periodic structural plates outside the elastic wave flexural band gap (12Hz) (a), and within the elastic wave flexural band gap (30 Hz) (b).

4.Conclusion

A plate with periodic cylindrical oscillators is proposed in this paper to realize the vibration isolation with a low frequency. Each oscillator consists of a small rubber cap and a large steel cap. By introducing the concept of band structure in solid mechanics, the band structure of a cell is analysed, and the range of flexural band gap and longitudinal band gap is concluded with the modal displacement diagram. The results are further verified with the calculation of transmission curves of the whole plate. Finally, the displacement modes of the plate are plotted. It shows that vibrations within the band gap are prohibited. When applied to some engineering structures or large mechanical equipment, the research results are of great significance for vibration reduction and isolation in low frequency domain. Further research should concentrate on how to achieve a lower and wider band gap with smaller oscillators through proper design.

5. References

[1] M. M. Sigalas and E. N. Economou, Elastic and Acoustic wave band structure. J. Sound Vib. 158, 377 (1992).

[2] J. J. Chen, X. Han, and G. Y. Li, Asymmetric Lamb wave propagation in phononic crystal slabs with graded grating J. Appl. Phys. 113, 184506 (2013).

[3] H. Zhang, J. Chen, and X. Han, Lamb wave band gaps in a homogenous plate with periodic tapered surface. J. Appl. Phys. 112, 054503 (2012).

[4] Vasseur J O, Deymier P A, Chenni B, et al. Experimental and theoretical evidence for the existence of absolute acoustic band gaps in two-dimensional solid phononic crystals. J. Physical Review Letters, 86, 14 (2001).

[5] Z. Liu, X. Zhang, Y. Mao, Y. Y. Zhu, Z. Yang, C. T. Chan, and P. Sheng, Locally Resonant Sonic Materials.Science 289, 1734 (2000).

[6] Y. Pennec, B. Djafari-Rouhani, H. Larabi, J. O. Vasseur, and A. C. Hladky-Hennion, Low-frequency gaps in a phononic crystal constituted of cylindrical dots deposited on a thin homogeneous plate Phys. Rev. B 78, 104105 (2008).

[7] J. Hsu, Local resonances-induced low-frequency band gaps in two-dimensional phononic crystal slabs with periodic stepped resonators. J. Phys. D: Appl. Phys. 44, 055401 (2011).

[8] H.J. Zhao, H.W. Guo, B.Y Li, Z.Q. Deng, and R.Q. Liu. Flexural vibration band gaps in a double-side phononic crystal plate. Journal of Applied Physics. (2015).

[9] H.J. Zhao, H.W. Guo, M-Xing, R.Q. Liu and Z.Q. Deng. Vibration band gaps in double-vibrator pillared phononic crystal plate. Journal of Applied Physics. 119. 014903. 10.1063/1.4939484. (2016).

IOP Conf. Series: Materials Science and Engineering **831** (2020) 012006 doi:10.1088/1757-899X/831/1/012006

Experimental Investigation of Heterogeneous Strain Fields Created in Metallic Materials

Bashir S Shariat

Department of Mechanical Engineering, The University of Western Australia
35 Stirling Highway, Perth, WA 6009, Australia.

bashir.samsamshariat@uwa.edu.au

Abstract. This study reports experimental evaluation of inhomogeneous strain fields created in NiTi structures during stress-induced martensitic transformation. A superelastic NiTi thin plate was used to create samples with uniform and non-uniform geometries. Digital image correlation technique was employed to measure the local strain variation in the NiTi samples under mechanical loading. Clear Lüders band formation and propagation were observed during experiments. The local strain contours revealed that the deformation was localised and spatially inhomogeneous in all samples. In the geometrically non-uniform sample, the transformation initiation and propagation were controlled by geometrical gradient. Also, the global deformation behaviour exhibited stress-gradient during forward and reverse transformations because of geometrical non-uniformity.

1. Introduction

Because of their martensitic transformations, shape memory alloys (SMAs) show unique thermomechanical behaviours, most remarkably pseudoelasticity and shape memory effect [1, 2]. Pseudoelasticity is denoted as the recovery of deformation well beyond the elastic limit of metals spontaneously upon unloading. Shape memory effect is denoted as the recovery of that large deformation upon heating. Owing to these distinct features, they have been used in varied engineering applications, such as sensors, actuators, and medical devices [3, 4]. The most broadly used SMAs are near-equiatomic NiTi alloys, which undertake B2↔B19′ martensitic transformations, where B2 is the austenite (A) and B19′ is the martensite (M) [5]. The A↔M transformations can be induced by application of either stress or temperature change. During stress-induced transformations, NiTi typically experiences a Lüders-like deformation, which yields in a stress plateau over the transformation strain of 6~8% [6, 7].

The deformation behaviour of SMAs including NiTi is very different from that of conventional metallic materials due to the participation of martensitic transformation. A generic metal displays typically a smooth stress-strain curve involving elastic and plastic deformations. In contrast, NiTi alloy exhibits a large stress plateau in a Lüders-like manner prior to proceeding to the more conventional elastoplastic deformation similar to that of a generic metal. This stress plateau, which is also sensitive to temperature, provides a discontinuity in the stress-strain curve and represents a case of mechanical instability with a zero strain hardening coefficient. Because of Lüders band formation and propagation, the deformation of NiTi is localised and spatially inhomogeneous during tensile loading, which creates heterogeneous strain field within the structure upon loading. In the case of geometrical non-uniformity, the variation in geometry causes the structure to experience an inhomogeneous stress field upon loading, thus progressive martensitic transformation occurs within the structure. This increases the level of complexity and inhomogeneity of the transformation field and the local strain field within the structure [8, 9]. This phenomenon has attracted much attention for researchers to develop theoretical understanding of the deformation behaviour of SMAs with geometrical non-uniformity [10, 11]. The geometrical non-uniformity can be created by regular variation of geometry as in geometrically graded structures [12, 13] or by irregular variation of geometry as in perforated or porous structures [14, 15]. In geometrically graded NiTi structures, the propagation of transformation is affected by the geometrical gradient direction [16, 17]. It has been reported that the directional transformation propagation can also be created by microstructural or compositional gradients within NiTi structures [18, 19].

This article presents an experimental investigation of the heterogeneous strain fields created in NiTi specimens with uniform and non-uniform geometries during tension. Digital image correlation (DIC) technique was used to capture the Lüders band formation and propagation and to obtain the strain field variation within the NiTi samples during tensile loading.

2. Materials and methods

A Ti-50.8at%Ni plate of 0.12 mm thickness was used to create geometrically uniform and non-uniform samples. The samples were fabricated by means of electrical discharge machining. Before each tensile experiment, a stochastic pattern was created on the surface of the sample by black and white paints for DIC measurement. Tensile experiments of the NiTi samples were conducted using an Instron 5982 machine with a slow strain rate of ~5×10^{-4}/sec. DIC method was applied to acquire local strain fields during tensile deformation. The testing temperature was 294 K.

In this paper, the nominal stress (σ) is defined as the load over the initial maximum cross-sectional area of the sample. Also, the nominal strain (ε) is defined as the elongation over the initial gauge length of the sample. For a geometrically graded sample, the width aspect ratio is defined as $\alpha = w_1/w_2$, where w_1 and w_2 are the minimum and the maximum widths, respectively.

3. Results and discussion

3.1. Uniform geometry

Here, the deformation behaviour of a uniform NiTi strip of 30 mm in gauge length and 2 mm in width under tension is presented. Figure 1(a) shows the global deformation behaviour acquired from the tensile testing machine. The sample was loaded up to 0.07 nominal strain and then unloaded. As observed, the material exhibited good pseudoelastic behaviour and recovery of deformation with a flat stress plateau at ~350 MPa over the forward A→M transformation. Figure 1(b) shows the variation of local strain field (ε_x) of the above sample during loading at progressive nominal strain levels, which was obtained through DIC. As observed, multiple Lüders band formed and the material exhibited inhomogeneous strain field along the x-axis during forward transformation.

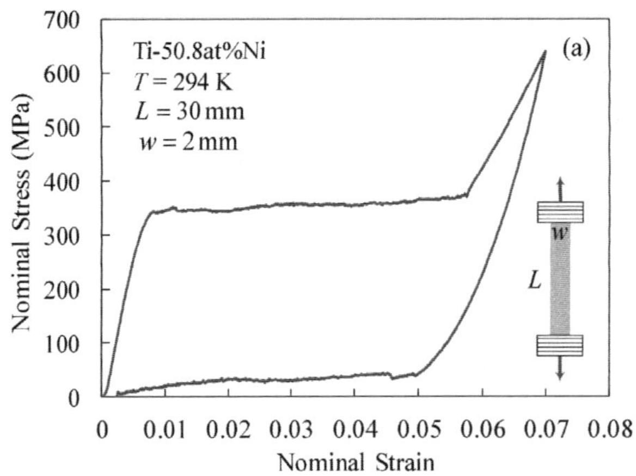

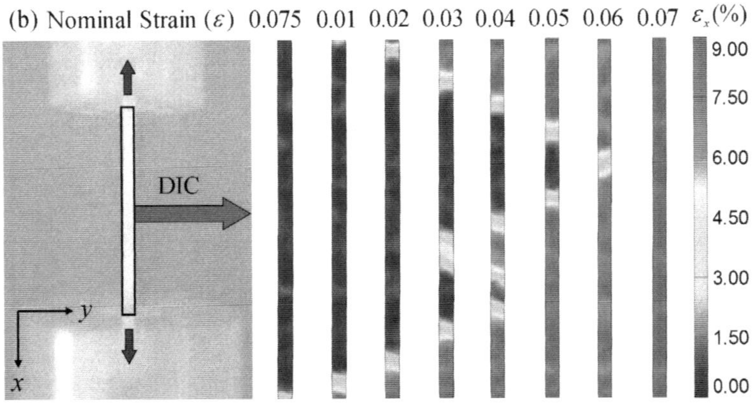

Figure 1. The deformation behaviour of a NiTi structure with uniform geometry under tension; (a): global deformation behaviour, (b): local deformation behaviour.

3.2. Non-uniform geometry

Here, the deformation behaviour of a geometrically graded NiTi strip under tension is presented. The sample was tapered linearly along the gauge length of 30 mm. The maximum width was 6 mm. The width aspect ratio was 0.6. Figure 2 (a) shows the global deformation behaviour acquired from the tensile testing machine. The sample was loaded up to 0.07 nominal strain and then unloaded. As observed, the material exhibited stress gradient over A↔M transformations. Figure 2(b) shows the variation of local strain field (ε_x) during tensile loading at progressive nominal strain levels, which was obtained through DIC. The strain contours reveal heterogeneous strain field within the geometrically graded structure along the *x*-direction during forward transformation. As observed, the A→M transformation initiated at the top end with the minimum width, where the stress was maximum. Then, the transformation progressively propagated toward the bottom end with the maximum width by the increase of load. This progressive martensitic transformation, which was stimulated by geometrical variation, created stress gradient over transformation as seen in figure 2(a). This experimental observation has been recently modelled by finite element method [20].

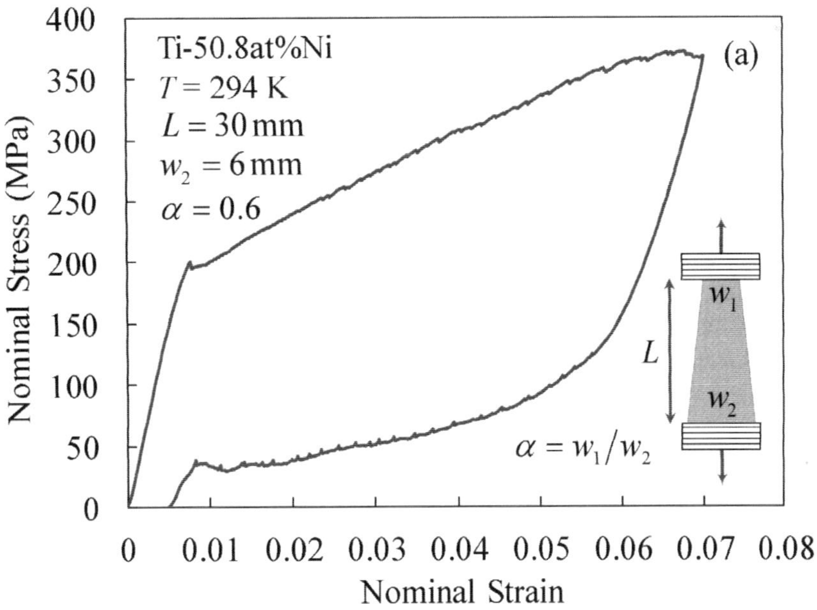

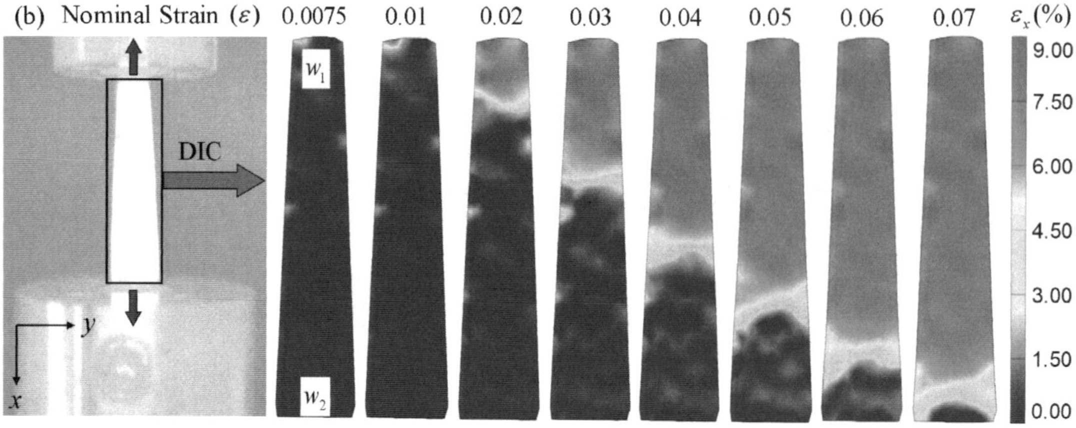

Figure 2. The deformation behaviour of a NiTi structure with non-uniform geometry under tension; (a): global deformation behaviour, (b): local deformation behaviour.

4. Conclusion

In this study, the local strain field of NiTi alloy has been investigated through digital image correlation technique. Heterogeneous strain fields were observed in NiTi structures with uniform and non-uniform geometries. This heterogeneity in strain field was due to Lüders-like deformation behaviour of NiTi and also progressive martensitic transformation due to geometrical non-uniformity. In non-uniform NiTi structures, the transformation initiation and propagation were controlled by geometry. In a NiTi strip tapered along the length, the stress-induced transformation initiated from the narrower end and propagated toward the wider end. The global stress-strain curve exhibited stress gradient over forward and reverse transformations.

5. References

[1] Otsuka, K. and X. Ren, Progress in Materials Science, 2005. **50**(5): p. 511-678.
[2] Shariat, B.S., Y. Liu, and G. Rio, Intermetallics, 2014. **50**: p. 59-64.
[3] Mohd Jani, J., et al., Materials & Design, 2014. **56**: p. 1078-1113.
[4] Sun, L., et al., Materials & Design, 2012. **33**: p. 577-640.
[5] Bakhtiari, S.R., et al., Scripta Materialia, 2018. **151**: p. 57-60.
[6] Hallai, J.F. and S. Kyriakides, International Journal of Plasticity, 2013. **47**: p. 1-12.
[7] Shariat, B.S., et al., Materials & Design, 2017. **124**: p. 225-237.
[8] Bakhtiari, R., et al., Functional Materials Letters, 2017. **10**(01): p. 1740011.
[9] Shariat, B.S., R. Bakhtiari, and Y. Liu, J of Alloys and Compounds, 2019. **774**: p. 1260-1266.
[10] Liu, H., J. Wang, and H.-H. Dai, Mechanics of Materials, 2017. **112**: p. 40-55.
[11] Shariat, B.S., Y. Liu, and S. Bakhtiari, J of Alloys and Compounds, 2019. **791**: p. 711-721.
[12] Shariat, B.S., Y. Liu, and G. Rio, Materials & Design, 2013. **50**: p. 879-885.
[13] Shariat, B.S., Y. Liu, and G. Rio, J of Alloys and Compounds, 2013. **577, Supplement 1**: p. S76-S82.
[14] S. Shariat, B., Y. Liu, and G. Rio, Materials Science Forum, 2010. **654-656**: p. 2091-2094.
[15] Shariat, B.S., Y. Liu, and G. Rio, Journal of Intelligent Material Systems and Structures, 2014. **25**(12): p. 1445-1455.
[16] Shariat, B.S., et al., Journal of Alloys and Compounds, 2019. **806**: p. 1522-1528.
[17] Shariat, B.S., et al., Data in Brief, 2019. **27**: p. 104566.
[18] Meng, Q., et al., Scripta Materialia, 2017. **127**: p. 84-87.
[19] Shariat, B.S., et al., Acta Materialia, 2013. **61**(9): p. 3411-3421.
[20] Shariat, B.S., S. Bakhtiari, and Y. Liu, IOP Conference Series: Materials Science and Engineering, 2019. **522**: p. 012005.

IOP Conf. Series: Materials Science and Engineering **831** (2020) 012006 doi:10.1088/1757-899X/831/1/012006

Acknowledgments
We wish to acknowledge the financial support to this work from the Australian Research Council in grant DE150101795.

IOP Conf. Series: Materials Science and Engineering **831** (2020) 012007 doi:10.1088/1757-899X/831/1/012007

A Numerical Investigation into the Impact of Icing on the Aerodynamic Performance of Aerofoils

Dinesh Bhatia[1,a], **Al Yaqadan Shaaban al Wahaibi**[2]

[1]The University of Nottingham Ningbo China, School of Aerospace, 199 Taikang E. Road, Ningbo, China
[2]Military Technological College, Department of Aeronautical Engineering, 262 Al Matar Street, Muscat, Sultanate of Oman

[a] Corresponding author: Dinesh.Bhatia@nottingham.edu.cn

Abstract. Aircraft wings and wind turbine blades are often subjected to harsh and cold climatic conditions. Icing is often observed on wing and blade surfaces in these cold climatic conditions. Wind turbine blades, in particular, are severely impacted by ice accretion which greatly hampers their performance and energy generation efficiency. Ice-accretion patterns are observed to vary with changes in temperature. As the temperature changes, the thickness of the ice accretion, the shape and location of ice-accretion vary greatly. In this paper, three different ice accretion patterns and their impact on the aerofoil efficiency have been investigated using the SST $k-\omega$ model in ANSYS CFD. An analysis of the impact of ice-accretion through a comparison of lift and drag coefficients for all three ice accretion patterns indicate that the accretion of ice on an aerofoil can reduce lift generation by 75.3% and increase drag by 280% thereby severely impacting the performance of the aerofoil. The loss in aerodynamic performance is greatly dependent on the ridge height, the extent of ice accretion and the thickness of this ice. The loss in aerodynamic performance has no fixed correlation to the drop in temperature.

1. Introduction

Air-transportation and renewable energy generation has seen an exponential increase in their growth over the last two decades. With climate change being an ever-increasing reality, energy efficiency is the key to reduce global carbon emissions. Reduction in emissions from aircraft can be through improved aerodynamic performance and reduction in drag. In the case of wind turbines, a reduction in downtime can improve the efficiency of energy generation.

Aircraft are increasingly travelling to new destinations some of which are subject to extremely cold climatic conditions. Similarly, wind turbine installations are increasingly taking place in regions such as Greenland, Norway, Denmark, and Canada etc. which often face harsh winters. It is estimated that 24% of the installed wind power capacity is in cold regions where temperatures are often below 0°C [1]. Icing has a very serious impact on the safety of aircraft and loss of power generation in the case of wind turbines. 240 cases of aviation accidents have been reported from 1990-2005 due to the presence of icing [2], [3]. Icing has also led to a 16% decline in the power generation capability of installed wind turbines [4]. Thus, it is imperative that the phenomenon of icing should be studied and remedial action taken to improve efficiency and safety of aircraft and wind turbines.

Icing on aerofoils has been studied extensively with a number ice-accretion experiments being conducted, details of which have been provided by Pouryussefi et al. [3]. One of the key outcomes from

IOP Conf. Series: Materials Science and Engineering **831** (2020) 012007 doi:10.1088/1757-899X/831/1/012007

experimental testing of ice-accretion was the classification of the type of ice as well as the impact of non-dimensional ice height (k/c) where k is the height/thickness of the ice accretion and c-chord length of the aerofoil. Ice can be classified into 5 main types based on the their shape: (1) roughness ice, (2) horn ice, (3) streamwise ice, (4) spanwise ridge ice, and (5) runback ice [5]. Wind tunnel tests indicate that presence of icing can reduce the lift coefficient within a range of 50- 60% as compared to a clean aerofoil although the data for the increase in drag coefficient is not readily available [3], [5]. While the impact of ice-accretion has been quantified experimentally, the use of simulations to quantify this effect has been few and far between. Simulations have usually been conducted using LES and DNS and user defined CFD models which make them time-consuming.

The purpose of this paper is to quantify the effects of three classical ice-shapes namely Ice Shape 290, 296 and 312 accurately, reliably and quickly on the lift and drag of a NACA23012 aerofoil numerically using a commercial CFD software (ANSYS). Details of these ice shapes are readily available in literature [5]. The impact of the ice height and location will also be studied.

This paper aims to demonstrate the viability of using a commercial CFD software along with a readily available CFD model (the SST $k-\omega$ model) to ensure rapid and accurate results for quantifying the impact of ice accretion on aerofoils. The paper will also provide the basic reference for any future study aimed at quantifying the impact different ice-accretion scenarios on the aerofoil performance.

2. Methodology

2.1. Mesh Generation

2D simulations were conducted and results were computed using the $k-\omega\ SST$ Turbulence model in ANSYS FLUENT. The meshing was carried out using the default workbench meshing software in ANSYS. NACA 23012 was chosen as the default aerofoil for the simulations and ice accretion profiles were digitised based on profiles obtained from existing wind tunnel ice accretion tests by Broeren et al. [5] as shown in Figure 1.

The aerofoil is assumed to have a chord length of 1.0 metre with the origin at the leading edge (0,0,0) and the trailing edge to be at (1,0,0). The meshing criteria was selected based on the mesh parameters set by Parashar [6].

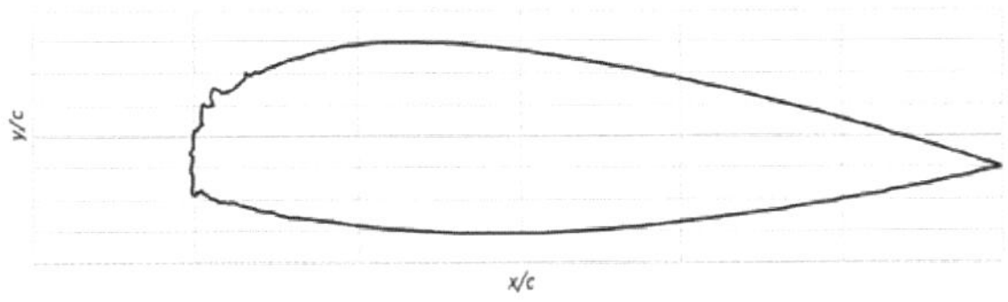

Figure 1. Digitised Ice Shape 290 from existing experimental data

The computational domain was set to be 20c (c-chord length of the aerofoil) upstream and downstream. A no-slip wall boundary condition was used on all solid boundaries whereas a farfield condition was set for the upper and lower boundaries. The boundary on the left was set as a velocity inlet whereas the one on the right was set to be a pressure outlet. A hybrid grid mesh was used with a $y^{+}=1.0$ near the aerofoil walls and a wall expansion ratio of 1.07. A mesh-independence check was conducted, and it was found that a mesh with 59,000 elements produced a truly mesh-independence solution.

Figure 2 shows an example of a mesh around a digitised iced aerofoil with a structured mesh around the aerofoil to resolve the boundary layer and an unstructured mesh further away from aerofoil.

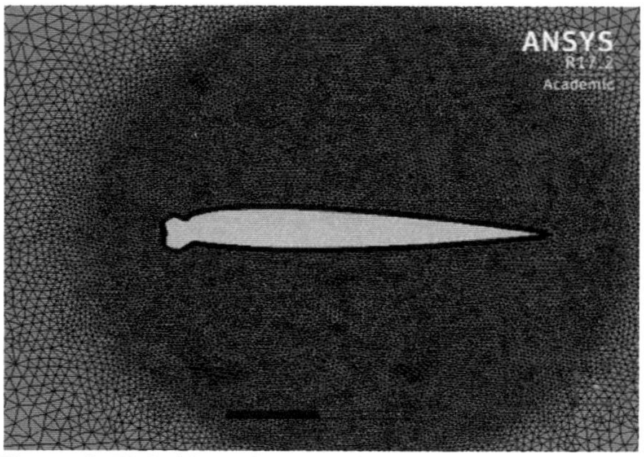

Figure 2. Hybrid mesh around a digitised iced aerofoil

2.2. Validation

Validation of the simulations were conducted against experimental tests done by Pouryoussefi et al. at a Reynolds number of 0.6×10^6 keeping all other parameters constant [3]. Plots of the lift coefficient C_l vs. the angle of attack α and the drag coefficient C_d vs. the angle of attack α were plotted for the NACA23012 aerofoil to obtain good correlation between the simulation method and experimental data. The plots are shown in Figure 3 below.

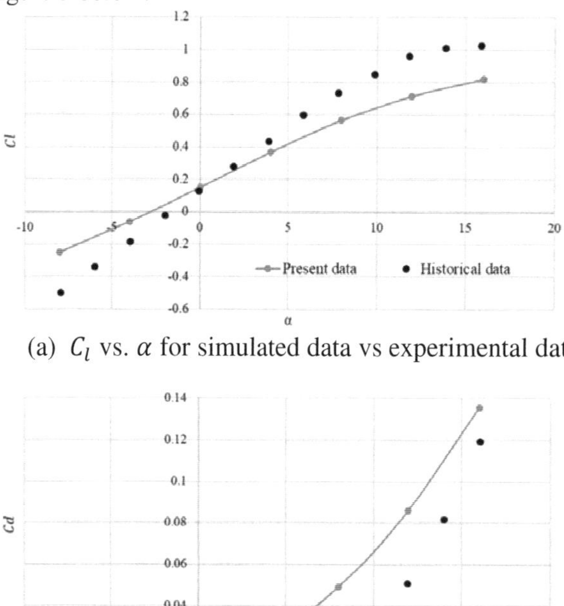

(a) C_l vs. α for simulated data vs experimental data

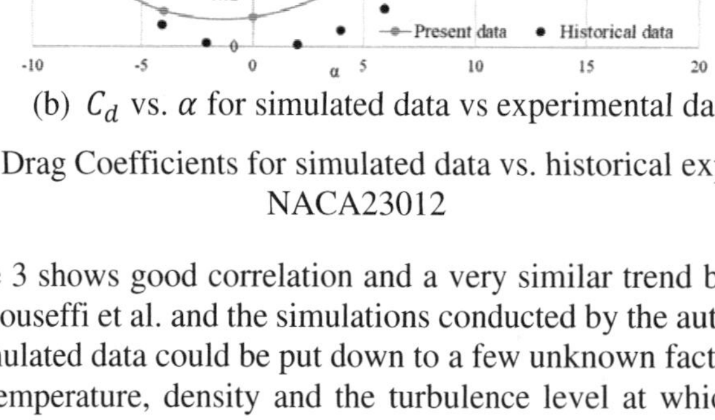

(b) C_d vs. α for simulated data vs experimental data

Figure 3. Lift and Drag Coefficients for simulated data vs. historical experimental data for NACA23012

Data presented in Figure 3 shows good correlation and a very similar trend between the experimental tests conducted by Pouryouseffi et al. and the simulations conducted by the authors. Variations between the experimental and simulated data could be put down to a few unknown factors such as the use of 2D simulations as well as temperature, density and the turbulence level at which the experiments were conducted. Temperature, density and turbulence were kept at default values during the simulations by the authors.

3. Results and Discussion

Three classical icing scenarios were considered in order to analyse their aerodynamic performance vs. the clean aerofoil. Ice Shape 290, 296 and 312 were considered based on the type of ice, location of the ice ridge on the aerofoil along with the height of the ice ridge.
Table **1** provides specific details about these ice shapes whereas Figure 4 shows the specific shapes from existing experimental data [7].

Table 1. Test conditions and specifications for different ice-shapes

Ice-Shape	T (K)	Ridge Height (k/c)	Location (x/c)
290	263	0.0138	0.020
296	268	0.0122	0.035
312	258	0.0131	0.060

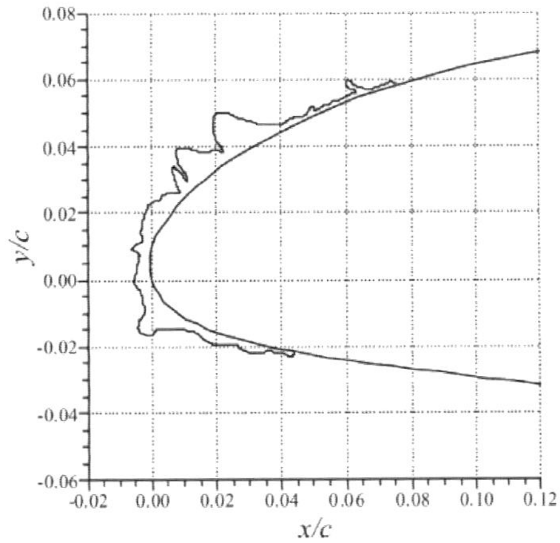

(a) Ice-Shape 290

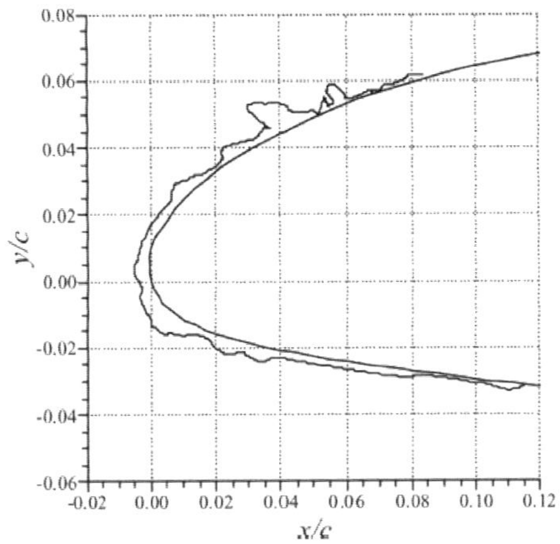

(b) Ice-Shape 296

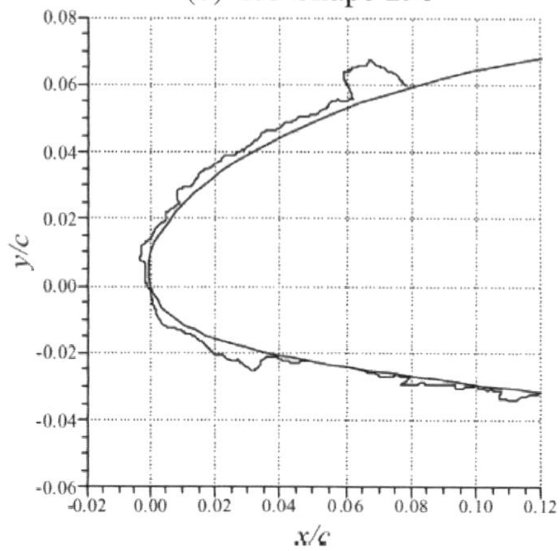

(c) Ice-Shape 312

Figure 4. Ice Shapes 290, 296 and 312 depicting different ice accretion patterns [7]

All three ice-shapes represent vastly different locations for ice-accretion based on the temperature. The temperature also impacts the type of ice that accumulates on the aerofoil. Ice-shape 290 represents a mixed glaze-rime shape which leads to substantial ice-accumulation closer to leading edge. Ice-shape 296 and 312 are glaze shaped ice with different locations and height for the ridge shaped ice accretions. The presence of these different ridge shaped ice-accretions greatly impacts the aerodynamic performance of the aerofoils.

A comparison of the lift coefficient C_l vs. the angle of attack α (shown in Figure 5) indicates that the ice-shapes 296 and 312 have very similar performance, but show a 60% reduction in lift as compared to the clean aerofoil. On the other hand, ice-shape 290 has the greatest lift reduction of about 75% as compared to the clean aerofoil. A very crucial factor that affects the performance is the height of ice-shape, with an increase in the height greatly reducing the lift coefficient (C_l). Ice-shape 290 had the largest ridge height of $k/c = 0.0138$ which resulted in significant drop in the lift coefficient (C_l). The lift coefficient (C_l) is marginally higher for ice-shape 312 which can be attributed to the fact the ridge of the ice is located the furthest away from the leading edge at $({}^{x}/_{c}) = 0.060$.

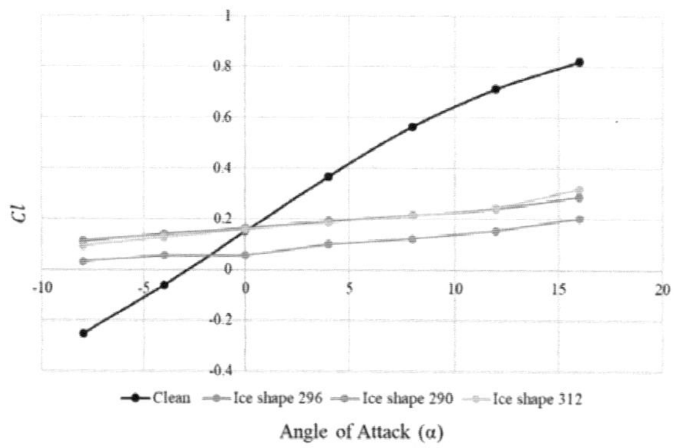

Figure 5. Lift coefficient comparison for three different ice-shapes vs. the clean aerofoil

A comparison of the drag coefficient (C_d) vs. the angle of attack α (shown in Figure 6) indicates that ice-shape 312 has the highest drag as compared to the clean aerofoil. Ice-shape 290 has the highest ridge height but the lowest extend of ice accretion around the upper and lower surfaces of the aerofoil whereas ice shape 296 has the lowest ridge height but the extent of ice accretion around the aerofoil extends upto $({}^{x}/_{c}) = 0.11$ on the lower surface albeit with an overall lower thickness as compared to ice-shape 290.

This possibly results in the lowest drag amongst the three ice-shapes. In the case of ice-shape 312, the ridge is located the furthest along the aerofoil at $(^{x}/_{c}) = 0.060$, the extent of the ice accretion is the furthest until $(^{x}/_{c}) = 0.12$ on the lower surface. A combination of the second highest ridge height and the largetst extent of ice-accretion along the aerofoil in the streamwise direction might have contributed to a higher drag for ice-shape 312 as compared to the clean aerofoil. In the case ice-shape 312, the drag is 280% higher than in the case of the clean aerofoil thereby indicating a significant loss in aerodynamic performance.

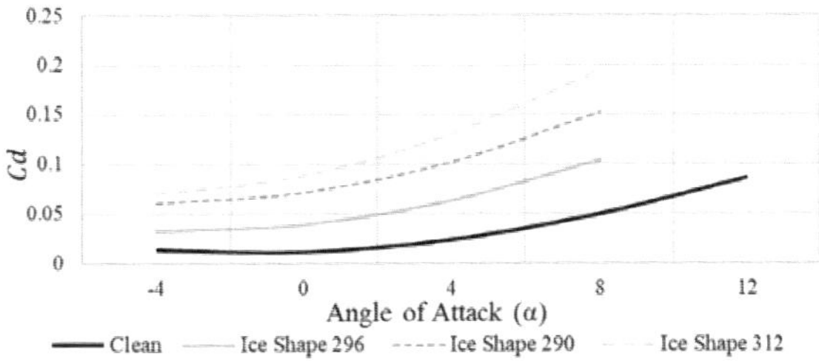

Figure 6. Drag coefficient comparison for three different ice-shapes vs. the clean aerofoil

Table 2 Summarises the relative performance of the aerofoil for the three different ice shapes vs the clean aerofoil.

Finally, a comparison between the boundary layer thickness of Ice shape 290 vs. the clean aerofoil was conducted as shown in Figure 7 and highlights the impact that ice accretion has on the overall flow structure. Ice shape 290 was used as it had the highest ridge height but the lowest overall ice extent over the aerofoil. For Ice Shape 290, the maximum ridge height is at $x/c = 0.08$. At this point, a comparison of the boundary layer thickness between Ice Shape 290 and the clean aerofoil indicates that the boundary layer for Ice Shape 290 is about 60% thicker which indicates a less streamlined fluid flow. The boundary layer thickness beyond the ice accretion area at $x/c = 0.2$ tends to show similar thickness. At this point, the Ice Shape 290 has an 8.6% thicker boundary layer. However, along the entire length of the aerofoil, Ice Shape 290 tends to depict a much thicker boundary layer as compared to the clean aerofoil which is a clear indicator of a less streamlined flow.

Table 2. Relative performance comparison between the iced-aerofoil and the clean aerofoil

Ice-Shape	Lift Decrease (%)	Drag Increase (%)
290	75.3	200
296	64.8	120
312	60.9	280

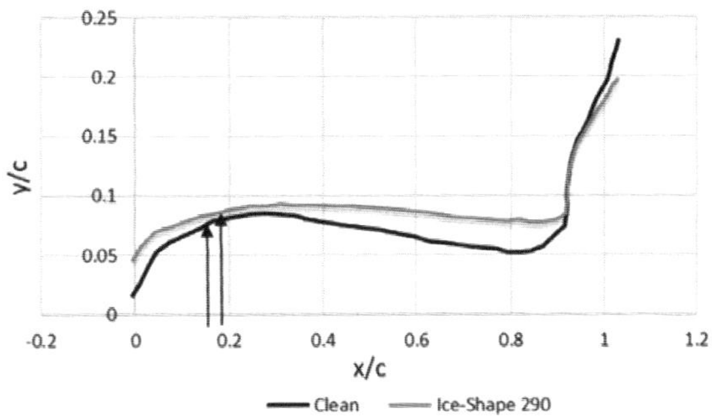

Figure 7. Boundary layer thickness comparison between clean aerofoil and Ice Shape 290

4. Conclusions

The impact of ice-accretion on aerofoils was studied numerically. The aerodynamic performance of three different classical ice accretion profiles namely ice-shapes 290, 296 and 312 were simulated using the SST $k - \omega$ model within a commercial CFD software and their results were compared to a clean NACA23012 aerofoil. Results indicate that the aerodynamic performance of an iced aerofoil is greatly impacted by the ridge height (k/c), the extent of ice accretion on the aerofoil surface as well as the thickness of this ice accretion. A larger concentration of ice around the leading edge (ice-shape 290) leads to a significant drop in the lift but doesn't necessarily lead to the largest increase in drag. However, the impact of the accumulated ice is observed along the entire length of the aerofoil and manifests itself in the form of a thicker boundary layer. A combination of the extended accumulation of ice (as seen in ice-shape 312) along with the presence of a ridge at the end of the ice accretion region leads to a significant increase in drag.

There is no fixed correlation between a drop in the temperature and the loss of aerodynamic performance. The substantial increase in drag and the drop in lift has great repercussions for the smooth operations of aircraft as well as wind turbines. Effective solutions are essential for anti-icing of the aerofoil or to de-ice the aerofoil surface once the ice has accumulated without significant consumption of energy as that will greatly impact the overall efficiency of the aircraft or wind turbine

The use of the SST $k - \omega$ model and a commercial CFD software have shown to provide rapid, reliable and repeatable results that should enable quantification of performance loss in a wide range of ice accretion scenarios.

5. References

[1] A. Hudecz, "Icing Problems of Wind Turbine Blades in Cold Climates," Technical University of Denmark, 2014.

[2] J. Mason, W. Strapp, and P. Chow, "The Ice Particle Threat to Engines in Flight," in *44th AIAA Aerospace Sciences Meeting and Exhibit*, 2006.

[3] S. G. Pouryoussefi, M. Mirzaei, M.-M. Nazemi, M. Fouladi, and A. Doostmahmoudi, "Experimental study of ice accretion effects on aerodynamic performance of an NACA 23012 airfoil," *Chinese J. Aeronaut.*, vol. 29, no. 3, pp. 585–595, Jun. 2016.

[4] "Wind Energy in Cold Climates | Natural Resources Canada," *Natural Resources Canada*, 2017. [Online]. Available: https://www.nrcan.gc.ca/energy/energy-sources-distribution/renewables/wind-energy/wind-energy-cold-climates/7321. [Accessed: 19-Jul-2019].

[5] A. P. Broeren, E. A. Whalen, G. T. Busch, and M. B. Bragg, "Aerodynamic Simulation of Runback Ice Accretion," *J. Aircr.*, vol. 47, no. 3, pp. 924–939, May 2010.

[6] H. Parashar, "Calculation of Aerodynamic Characteristics of NACA 2415, 23012, 23015 Airfoils Using Computational Fluid Dynamics (CFD)," *Int. J. Sci. Eng. Technol. Res.*, vol. 4, no. 3, pp. 610–614, 2015.

[7] A. P. Broeren, M. B. Bragg, and H. E. Addy, "Effect of Intercycle Ice Accretions on Airfoil Performance," *J. Aircr.*, vol. 41, no. 1, 2004.

Intelligent Design Based on Knowledge Capture and Reuse

Hao Qin[1,a], Jiong Jin[2], Yu Zhang[1], Shilin Hu[3]

[1]Guangdong Institute of Intelligent Manufacturing, Guangzhou, China
[2]Swinburne University of Technology, Melbourne, Australia
[3]Kunming University of Science and Technology, Kunming, China

[a] h.qin@giim.ac.cn

Abstract. Engineering design is a creative activity which is depending on both scientific knowledge and innovative ideas of human being. Intelligently recommend the scientific knowledge to the designers would certainly shorten the time they spent in finding the data, information and knowledge they need for design decision-making. Thus, they can spend more time in the innovative activity of design. The design of a flexible gear for harmonic drive reducer is a knowledge intense job, where intelligent design method is required to facilitate the design process and thus well applicable. This paper explores a systematic methodology of undertaking intelligent design of such a mechanical component, through on knowledge capture, representation and reuse. A case study on a double-arc flexible gear design is undertaken as the evaluation and improvement of this methodology, whose results have shown the good effectiveness of the methodology in shortening the time for modelling, information retrieval and decision-making.

1. Introduction

Engineering design is a creative activity which is depending on both scientific knowledge and innovative ideas of human being. The creative ability is various for each designer, while is also depended on the data, information and knowledge he or she possessed. Since engineering designers spend above 50% of their time looking for data and information to assist their design tasks, finding a way to intelligently deliver the data and information to the designers needed is significant in improving design efficiency. This intelligent recommendation requires to firstly recognize the working context of the designers, and then provide the data and information related to the design context. In this way, the time engineering designers used in searching data and information is dramatically shorten, which allows the designers to spend more time on innovative part of the design. However, present computer-aided design tools focus either on streamlining the design procedures (e.g. 2D/3D modelling) or undertaking design documentation (e.g. Product Data Management), which lacks a kind of design support tool in intelligently recommending the possible data/information designers required for their decision-making on design tasks. To fill this gap, this paper proposes an intelligent design methodology on how to improve design efficiency through intelligent recommendation. This methodology includes three parts; the first one is to acquire the useful data, information and knowledge during the design process, the second one is to present them in a structured way, and the third one is to use intelligent Q&A system for potential solution recommendation. A design of flexible gear for harmonic drive reducer is chosen as example to demonstrate the methodology, where testing have been done to evaluate the effectiveness in undertaking design in a more intelligent way.

The rest of the paper is organized as follows. Section 2 reviews relevant work on design supporting methods and computer-aided design tools. Section 3 explains how to acquire the data, information and knowledge on design object and process, together with Section 4 describing how to present them in a systematic and structured way. Section 5 demonstrates the intelligent recommendation system depending on the design knowledge captured. A discussion and conclusion are drawn in Section 6.

2. Computer support tools for knowledge capture and reuse

Engineering design is a complex process where a range of data and information are quired for its decision-making. To support the design process, computer-aided tools are developed and used to assist capturing, managing and reusing the data, information and knowledge, which can be regarded as knowledge management tools for knowledge generation, codification, and transfer [1]. The traditional method for implementing knowledge management is to use Product Data Management (PDM) systems to store design data and information in databases, which merely provide access to schematics, Computer-Aided Design (CAD) models, and documentations[2]. PDM systems manage documents such as reports, CAD data, engineering drawings, and analysis data through the design process. However, there are several shortcomings of these PDM systems, such as the lack of formal product representation in the form of function, behaviour and structure, lack of reuse of design knowledge, lack of impact analysis, etc., [3]. Aiming to improve the traditional method, Product Lifecycle Management (PLM) systems have been developed. The concept of PLM appeared in the 1990s as an extension of PDM [4], and the PLM is defined as the process of managing a company's products from their conception, design, manufacturing, all the way to its use and disposal [5]. The purpose of PLM systems is to integrate information on the manufacturing processes (CAM systems) with design data (CAD systems) on the one hand, and information about Enterprise Resource Planning (ERP) processes on the other hand [6]. However, these PLM systems lack essential capability for the management and reuse of design knowledge [7], [8] and are less suited for the conceptual design stage [7], [9]. Moreover, a significant shortcoming of existing PLM systems is that they are lack of adequate information models for product presentation, as these models would be needed to effectively capture, exchange, retrieve, and reuse design knowledge [10].

Knowledge management system method is a systematic process of acquiring, accumulating, and sharing individual knowledge, and uses them to strengthen competiveness [11], and a knowledge management system should be able to explicitly represent the tacit knowledge of designers, and facilitate expansion of the knowledge through the knowledge acquiring-storing-using cycle [12]. Recent research is focusing on developing a generic representation of product knowledge that includes other kinds of product knowledge beyond structure, function, and behaviour, in order to support a broader level of information exchange and interoperability [13]. Product knowledge is represented in terms of requirements, specification, artefacts, structure, functions, behaviours, design rationale, constraints and relationships. Such representation supports multiple levels of abstraction, which provides computational support for early design activities [10]. The support tools currently available for capturing design rationale using visualised and structured knowledge representation include the Compendium [14], [15] and the Design Rationale Editor (DRed) [16]. Also, the DRed tool has been developed and successfully integrated to the Rolls-Royce PLM toolset [4], [16], which provides a solution for the integrated knowledge management in engineering design.

3. Knowledge capture on the design object and process

In order to intelligently recommend the data, information and knowledge to engineering designer, knowledge acquisition from previous design projects are firstly required. In an engineering design project, the knowledge on decision-making of a design task is the key for the intelligent recommendation. This kind of knowledge includes structured knowledge (explicit knowledge) and unstructured knowledge (tacit knowledge). The previous one is mainly from formula, methods and theories, which is easier to be captured. The later one is more likely from designers' experience embedded in their brains, which is difficult to be captured and reused. From the authors' previous research, creating a RFBSE knowledge representation model [17] for a specific design project is able to capture and represent both explicit and tacit knowledge.

In this research, a design on double-arc flexible gear for harmonic drive reducer is chosen as research subject, and the RFBSE model is used to undertake the knowledge acquisition task. During its engineering design process, the knowledge generated in five aspects are particular focused, which are the requirements of the design as well as how to transfer the general requirement into technical function of the design object, the functional analysis and functional decomposition, the expected behaviour generation and actual behaviour evaluation, the structure synthesis and decomposition, and the design evolution. The design knowledge is captured in the form of Know-What, Know-How and Know-Why in a tree structure, resulting a whole knowledge map describing the design knowledge generated together with the relationships during the whole design process, as shown in Figure 1.

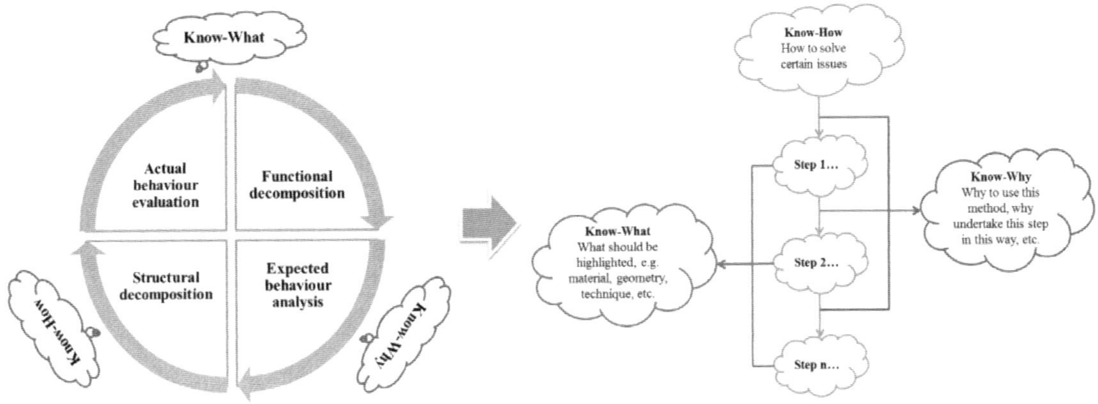

Figure 1. Method of Knowledge capture and representation

4. Representation of design knowledge

A Web-based knowledge management system is developed for the design knowledge representation, allowing engineering designers to easily record their design knowledge during the design process. The user interface of the system uses graphic representation to capture and represent the design knowledge in a tree structure. Firstly, a structural decomposion has been done to give a clear structure to the elements of the design object. Then, a range pieces of design knoweldge are attached to the 'branch' as 'leaves', where three-level represenatation is used, i.e. know-what, know-how, know-why. Among them, know-what describes the basic information of each element, know-how explains the process knowledge, while the know-why demonstrates the design rationale. By integating this three-level knowledge representaiton, the tacit knowledge embeded within engineering designer's brain can be captured in a systematic way for future reuse. Also, to give a clear view on design object, its 3D model has been attached alongside the knowledge representation tree, giving a 360 degree rotating view on the design object. as show in figure 2.

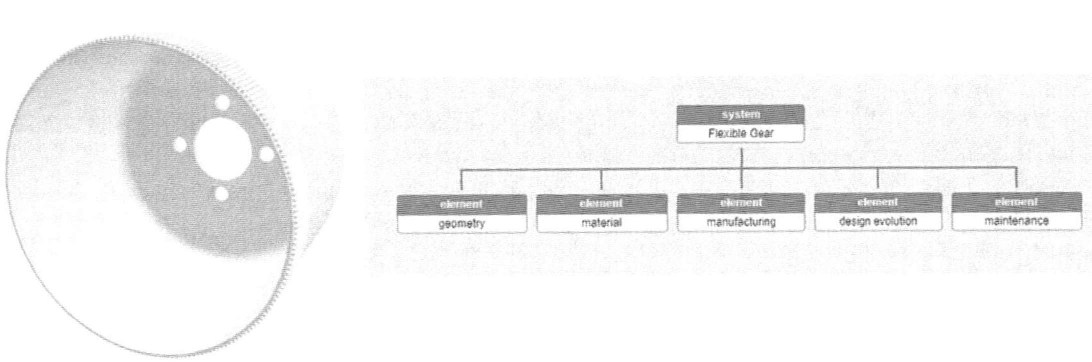

Figure 2. Knowledge capture and representation in the Web-based knowledge management system

In Figure 2, the tree structure on the right records the knowledge related to the design of the flexible gear, e.g. geometry, material, manufacturing, design evolution, maintenance, etc. More detail representation of these aspects can be added in terms of know-what, know-how and know-why, which gives an extendable frame for the knowledge capture and representation. On the right-hand side of the tree structure, a 3D model, i.e. a flexible gear in this example, has been displayed in panoramic view, which can give users a visual view on the design object and its design knowledge captured.

At the same time, a graphic database is used in the back- end to store the responding knowledge elements and their relationship. Also, any data and information related to the design object can be attached in respective element, forming a knowledge graph, as shown in Figure 3.

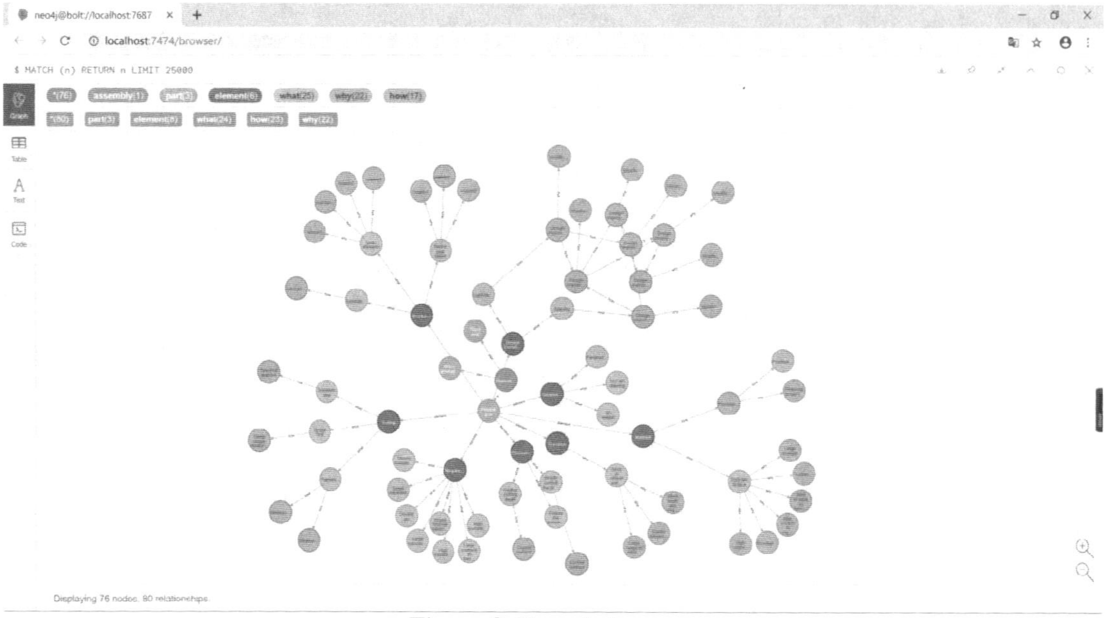

Figure 3. Knowledge map

Figure 3 shows how the knowledge elements are connected to each other. The assembly with orange colour can be decomposed into several parts with blue colour which is the minimum structure that cannot be decomposed anymore. For each part, there are several elements, among which know-what with brown colour is used to describe the basic information. Know-how with pink colour is used to explain the procedures while know- why with green colour is linked with know-what or know-how in demonstrating the design rationale. In this case, an integrated knowledge graph is created for knowledge representation and reuse.

5. Intelligent recommendation on the design solution

Based on the Web-based knowledge management system and the knowledge graph created, the design knowledge can be reused through intelligent recommendation according to the various working context of the engineering designers. When the designers use the system for recording design knowledge or retrieving specific data and information, the system will record their trace of use and then subsequently recognize the working context of the designers. For instance, when a designer is searching a material, the system will recognize he/she is at the stage of material selection, and then intelligently recommend several possible materials. Thanks to the knowledge representation model and knowledge graph, the relevant data and information can be easily and efficiently located in the database. For example, when a structure decomposition model of the robotic arm is created, it records the relationship between the flexible gear with the motor, harmonic drive reducer and robotic arm, the data and information of the flexible gear can be easily retried and recommended when designing a flexible gear in a robotic arm when the system has recognised the designer is designing a specific robotic arm, as shown in Figure4.

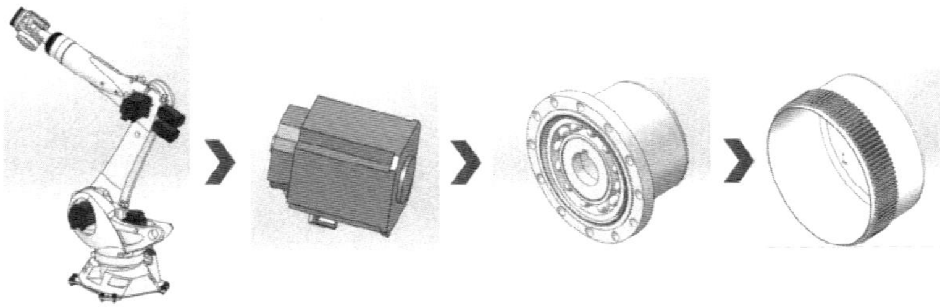

Figure 4. Locating specific data and information through intelligent recommendation

In Figure 4, the Web-based knowledge management system will firstly record the users' behaviours on searching the data, information and knowledge related to a robotic arm, the information they record in the system, and the key words or sentences they used for retrieving data and information, then the system can recognize the designer is designing a flexible gear and quickly located the relevant data and information in the database , and intelligently recommends the preferred information to the designer for their decision-making. In this way, the time that designers spent in finding relevant data and information they required for decision-making is significantly shorten, which eventually improves the design efficiency.

6. Conclusion

A large number of data and information are generated during engineering design process, and they are also required for design decision-making. In this case, efficiently finding and reusing these kinds of knowledge can significantly improve design efficiency through shortening the time in information retrieval. This paper has proposed an intelligent design method based on capturing and representing design knowledge, and subsequently providing them the designers based on their design context through intelligent recommendation. A case study of applying this method on a flexible gear design on harmonic drive for robotic arm has shown the feasibility of this method in improving design efficiency through knowledge capture and reuse.

7. References

[1] R. L. Ruggles, Knowledge management tools. 1997.
[2] S. Szykman, R. D. Sriram, C. Bochenek, J. W. Racz, and J. Senfaute, "Design repositories: engineering design's new knowledge base," IEEE Intell. Syst. their Appl., vol. 15, no. 3, 2000.
[3] T. Bilgic and D. Rock, "Product data management systems: State-of-the-art and the future," in Proceedings of the 1997 ASME design engineering technical conferences and computers in engineering conference, 1997.
[4] S. K. Chandrasegaran et al., "The evolution, challenges, and future of knowledge representation in product design systems," Comput. Des., vol. 45, no. 2, pp. 204–228, 2013.
[5] J. Stark, Product lifecycle management: 21st century paradigm for product realisation. 2005.
[6] S. Brandt, J. Morbach, M. Miatidis, M. Theisen, M. Jarke, and W. Marquardt, "An ontology-based approach to knowledge management in design processes," Comput. Chem. Eng., vol. 32, no. 1–2, pp. 320–342, Jan. 2008.
[7] J. Gao, H. Aziz, P. Maropoulos, and W. Cheung, "Application of product data management technologies for enterprise integration," International Journal of Computer Integrated Manufacturing, vol. 16, no. 7–8. pp. 491–500, 2003.
[8] P. Maropoulos, "Digital enterprise technology--defining perspectives and research priorities," Int. J. Comput. Integr. Manuf., vol. 16, no. 7–8, pp. 467–478, 2003.
[9] S. Mesihovic, J. Malmqvist, and P. Pikosz, "Product data management system-based support for engineering project management," Journal of Engineering Design, vol. 15, no. 4. pp. 389–403, 2004.
[10] S. Szykman, R. D. Sriram, and W. C. Regli, "The Role of Knowledge in Next-generation Product Development Systems," Journal of Computing and Information Science in Engineering, vol. 1, no. 1. p. 3, 2001.

[11] T. H. Davenport, D. W. De Long, and M. C. Beers, "Successful Knowledge Management Projects.," Sloan Manage. Rev., vol. 39, no. 2, pp. 43–57, 1998.
[12] J. Park, "Developing a knowledge management system for storing and using the design knowledge acquired in the process of a user-centered design of the next generation information appliances," Des. Stud., vol. 32, no. 5, pp. 482–513, 2011.
[13] J. Le Duigou, A. Bernard, N. Perry, and J. Delplace, "Inductive approach for the specification of a generic PLM system in an extended enterprise context.," 5th Int. Conf. Digit. Enterp. Technol., no. October, pp. 457–474, 2008.
[14] G. Conole, "Using Compendium as a Tool to Support the Design of Learning Activities," Knowl. Cartogr., p. 199, 2008.
[15] M. Shahin, P. Liang, and M. R. Khayyambashi, "Rationale visualization of software architectural design decision using compendium," Proc. 2010 ACM Symp. Appl. Comput. - SAC '10, p. 2367, 2010.
[16] R. Bracewell, K. Wallace, M. Moss, and D. Knott, "Capturing design rationale," Comput. Des., vol. 41, no. 3, pp. 173–186, 2009.
[17] H. Qin, H. Wang, and A. L. Johnson, "A RFBSE model for capturing engineers' useful knowledge and experience during the design process," Robot. Comput. Integr. Manuf., vol. 44, pp. 30–43, 2017.

Acknowledgments
Acknowledgements are given to the colleagues in Guangdong Intelligent Manufacturing Research Institute, and the financial support from Guangdong Academy of Science for the Special Fund of Introducing Doctoral Talent (Grand No. 2018GDASCX-0956) and International Science and Technology Cooperation Project (2019GDASYL-0503008).

IOP Conf. Series: Materials Science and Engineering **831** (2020) 012009 doi:10.1088/1757-899X/831/1/012009

Research on Design and Application of Automatic Production Line for Mechanical Components

Wu Bin

Guangzhou Railway Polytechnic, School of Mechanical and Electrical Engineering, 510430 Guangzhou, China

wulintian@126.com

Abstract. In order to meet the requirements of automatic processing for mechanical components manufacturing enterprises, the paper describes a design method of automatic production line. Firstly, the paper made the overall design of the automatic production line, takt time is calculated according to the annual production demand, and the analysis of device configuration and position layout of the production line was carried out based on the annual production requirement. Secondly, the paper selects types of CNC machines and industrial robots, and design structural of the silo, the temporary storage platform, the conveyor belt, the fuel injection and water removal. 3D model of each component and space layout of the production line are drawn by Solidworks software. Thirdly, the process flow of automatic production line was designed, and the CNC intelligent management system was developed to meet the requirements, which can monitor the process information such as takt time. Test is carried out after establishment of automatic production line, the test run finally shows that the automatic production line works stably and can meet the requirements of enterprise.

1. Introduction

With the development of automation technology and the gradual growth of labor costs, more and more enterprises pay attention to the automation of production processes. In order to improve the efficiency and stability of production, shorten production cycle and reduce product cost, automatic production lines are under establishment.

The automatic production line mainly includes CNC machine tools, industrial robots, fixtures, conveying equipment and so on. The PLC takes the general control to realize the automatic process of the entire production line. The paper gives the design and establishment of automatic production lines which take the actual processing needs of the company into consideration.

2. Design Method of automatic production line

The main works in designation of automatic production line include: the identification of takt time, the calculation of required work sites and equipment, the determination of the number of workers, the designation of the conveying method, the layout of production line, and the designation of production line process. After trial operation, make the comprehensive evaluation of the designed automatic production line [1].

First of all, the overall design scheme should be made, it is the most important part in designing and manufacturing process of the entire production line. The overall design scheme plays a decisive role in takt time, production efficiency, operational reliability, cost, designing and manufacturing cycle, which would determines whether the entire production line succeeds or not.

Secondly, select types of CNC machine tools and industrial robots, design the layout and PLC control systems of production lines. Based on the requirements of work site for robot, such as load, accuracy, working radius and speed, combine with the cost performance to select the type and model of the industrial robots.
Thirdly, the fixtures and various positioning devices are designed based on the structure and processing technology of products. The PLC control system designed to make the cooperation between the robot and other components smoothly to realize automatic production. Use the simulation software to measurement the effect of the entire automatic production line, and continuously optimize based on the results.
And then, put the assembly of all components through a test to complete the construction of the whole automatic production line.
There are some problems should be paid attention to in the designing process of the automatic line. For example, the related processes should be placed in a centralized manner to make the automatic production line run smoothly, the layout among all components should be put relatively compact to avoid crossovers in the process. To ensure takt time, the resource allocation and rate in the various processes of automatic production line should be controlled as balanced as possible, the layout should be also optimized to make full use of space, the future upgrade of the production line should be take into consideration, the interfaces needed to be reserved to make the production line flexible that can be suitable for many varieties production [2].

3. The overall design of the automatic production line

3.1 Analysis of automatic production line takt time

Takt time refers to the time that takes to produce one part on average from the automatic production line, we can optimize the production process and improve production efficiency by intergrating the automation line in the basis of desired takt time [3].
Production line takt time is calculated as

$$R=T/Q \tag{1}$$

In formula (1), R refers to the takt time, T refers to the planned effective working time, and Q refers to the planned product output.

$$T=T_1*K \tag{2}$$

In formula (2), T_1 refers to the planned working time, taking into account the impact of other factors on working hours, K is the working time utilization factor, and $K \leq 1$.
The takt time expected for the automation line can be calculated by the planned annual production and the effective working hours for enterprise, that means how much time one part takes would meet the annual output of the enterprise.

3.2 Overall configuration analysis

Based on the actual processing needs of enterprise, the automatic production line consists of two six-axis robots, four CNC machine tools, one set of three-in-one quality inspection stations, one set of two-in-one quality inspection stations, and one set of 32-position dot matrix fixed. The silo is composed of a conveying device and a dewatering equipment. Two industrial robots load and unload four CNC machine tools. Spot check every 10 parts processed, and the parts that are inspected are put into the sampling table by industrial robots. The silo is a pallet with a uniform structure to ensure the center position of the workpiece is consistent. The de-water injection equipment is equipped with a hot air system, which can quickly dry parts and is equipped with an automatic oil-passing system to make the workpiece covered with anti-rust oil. In addition, it is equipped with CNC intelligent management system, which can monitor the processing parameters of the production line. The screen can display the operating parameters of the equipment, processing capacity, such as basic data, equipment failure, number of workpiece processing, pass rate, takt time, downtime, etc. It's convenient for real-time monitoring and management.

4 Design of automatic production line components

4.1 Select the type of CNC and robot

The CNC machine tool is the machining center of automatic production line. When selecting the model, the factors such as comparison accuracy, spindle power and maximum spindle speed should be integrated. The practicality, economy, operability, stability and openness of the interface should also be considered. the vertical lathe was selected and the FANUC CNC operating system was adopted in the paper [4].

The industrial robot is an intelligent unit of automatic production line. It mainly undertakes functions such as loading and unloading, and its performance affects the processing quality of the workpiece and the takt time of the production line.

The maximum load, maximum working radius and degree of freedom of the robot are important factors to be considered in the selection of industrial robots. One industrial robot is loading and unloading for two CNC machine tools in this automatic production line, Selected 6-axis industrial robots based maximum rotation angle and working radius of each axis of the industrial robot., with a wrist load of 20KG and a forearm load of 10KG. The maximum rotation is shown in table 1 [5]. Figure 1 is the three-dimensional model of robot.

Table 1. Robot Joint Parameters

Joint	Joint Speed	Joint Angle
J1	170°/s	±165°
J2	170°/s	+155°~-90°
J3	152°/s	+75°~-200°
J4	350°/s	±360°
J5	333°/s	±120°
J6	600°/s	±360°

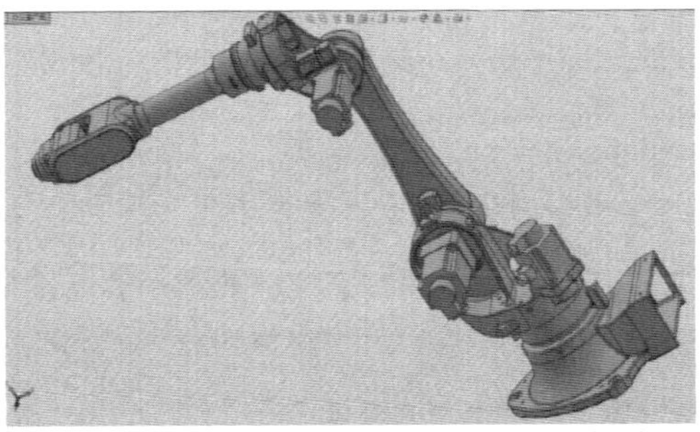

Figure 1. Three-Dimensional model of robot

The movement of the industrial robot is controlled by the teach pendant, and the basic motion of the robot is realized by interpolation commands such as MOVJ/MOVL. The robot loading and unloading program instruction is used as a subroutine. In the paper, the working mode of the robot is divided into manual mode and automatic mode. The robot in the manual mode can only manually control the robot to execute each motion program through the teach pendant. The robot in the automatic mode is triggered by the PLC to trigger the robot main program to call the relevant motion subroutine to complete the automatic operation of the robot.

4.2 Design of peripheral parts

The silo is a split double-layer structure, which can be placed with two layers of processed parts. When the processing of one side of the tray is completed, the handle can be pulled to move the silo out

of the working range of the robot, and the material is manually loaded again, so as to ensure personal safety and convenient to fill material.
The silo contains an inductive switch, and the alarm is working when the material on the tray is empty. There are two trays, 16 discs in a single disc, for a total of 64 workpieces.
When the robot changes the workpiece, four-station temporary storage stationstore the processed workpiece, wait for the mechanical hand material to be completed, and then pick up the processed workpiece for the next step.
The robotic gripper has a toothed adjustable structure for easy adjustment of specifications. The end of the robot is equipped with a laser detecting device. The detecting device is mounted on the hand to avoid contamination of the sensor by the cutting fluid of the machine. The sensor is placed outside of the machine during processing. The beating of the low-speed rotating blank on the machine tool is detected, the machine is closed when jump detection is OK, and the robot takes out the blank into the defective storage table when jump detection is NG. Figure 2 is three-dimensional structure model of robotic gripper .

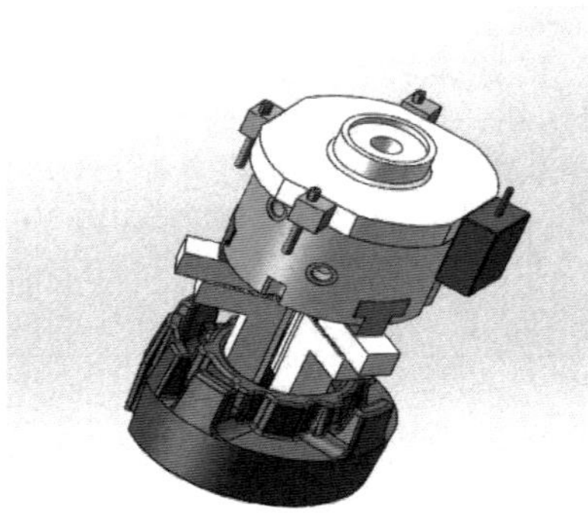

Figure 2. Structure model of robotic gripper

Two-in-one storage station is used for store the defective products and sampling products, it contains a sensor switch. when the workpiece is full, the alarm will work to prevent the robot from colliding.
The conveyor line is transported in the form of a double speed chain to meet the workpiece caching function, and the water such as cutting fluid can be dropped to the water tank to ensure the parts quickly remove moisture. It includes pusher mechanism, sensing device, 180° turning mechanism and filter cleaning tank.
Dewatering mechanism is used for drying workpieces, the first process uses air knife to form air curtain, the second process uses high pressure gas, and the third process uses hot air drying and evaporation, to ensure that the parts have no water residue, and then spray. It includes upper and lower fuel injection mechanisms, which have the fuel nozzles on the upper and lower layers of each of the six circumferences to ensure that the surface of the parts are covered with oil film the air knife generator and the movable baffle and so on. The fuel injection mechanism is installed before and after to prevent the oil mist from drifting out. After the baffle is lowered, the fuel injection starts, the fuel injection ends to open the baffle. Dewatering mechanism model as can be seen in Figure 3.

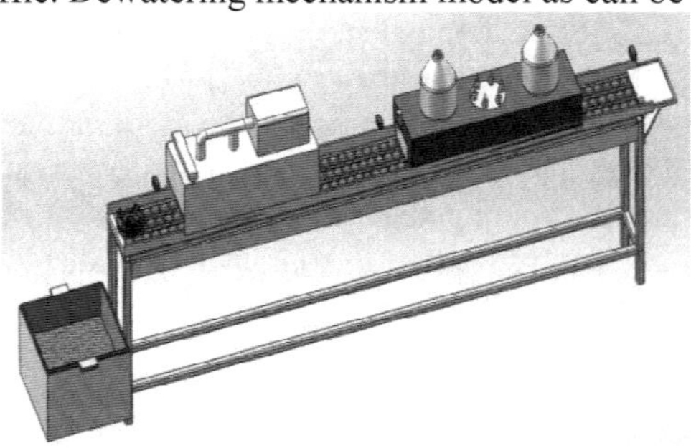

Figure 3. Dewatering mechanism model

4.3 CNC intelligent management system

The paper developed a CNC intelligent management system. The daily business volume of system is 10000 times each day, and the maximum response time is less than 2 seconds, daily inquiry is less than 10 seconds, report inquiry is less than 15 seconds, batch inquiry is less than 30 seconds.

The number of concurrent users is 50, and the maximum number of online users is 500. It can meet the actual needs of industrial site [6]. Table 2 is main functions of CNC management system.

Table 2. Main functions of management system

NO.	Module	Main Function Point
1	System management	system login, user management, rights management, log management, maintenance of various prompt information
2	Basic data management	Production line data, personnel configuration data, shift data maintenance; production work order data, equipment status data, etc.; equipment alarm data, equipment parameter abnormal data, equipment process slice analysis data, etc.
3	Status monitoring management	Real-time monitoring of different states such as power-on, power-off, alarm, and standby.
4	Production statisticsmanagement	Provides the corresponding processing quantity data under different program number conditions of the equipment.
5	Parameter monitoring management	Device parameter abnormal statistics, status monitoring, device parameter critical settings
6	Equipment Efficiency Analysis	Equipment CT and Equipment Availability UT Monitoring
7	Production slice analysis	Equipment time composition analysis, including different state slices such as fault, wait, edit, run, etc.
8	Production quantity statistics	Equipment output is based on the number of program numbers, and provides interfaces and work orders for docking.
9	Equipment fault management	Equipment fault code collection and time statistical analysis
10	DNC management	Device program preview, program upload, download, comparative analysis and other services
11	Interface	Receives automatic scanning of device data, receives status of processing equipment, receives tool status result determination of processing equipment, equipment processing result/device information system reception, spindle load reception

5. Processing technology of production line

The operator puts the blank material into the silo in turn, and the robot 1 grabs the billet from the silo and put into the CNC machine tool. The air gun on the robot blows the iron scrap on the tooling, then puts the blank material into the tooling, and the spindle rotates at a low speed to carry out runout detection, if runout detection is unqualified, the robot 1 moves the blank to the defective product temporary storage station, and re-takes the blank from the silo into the machine tooling. If the runout detection is qualified, it will be put into the machine tool 1 for processing, and the safety gate will be closed automatically by the PLC control, and CNC will start machining.

When the workpiece processed is finished, the safety door is opened automatically, the robot 1 takes out the workpiece, puts it into the cleaning pool 1 and shakes it cleanly, and the safety door closes automatically, the machine tool turret out of the chip cleaning tool, and then opens the safety door automatically.

1PC is sampled when 10PCS are processed, and the robot 1 moves the sampled product to the sampling station 1, and manually returns the transfer conveyor body after the inspection.

When the workpiece is cleaned, the robot 1 puts it into the transfer conveyor line for transmission and automatically completes the flip.

And then he robot 2 picks up the workpiece from the transfer conveyor and puts into the machine tool. The workpiece is close to the air gun blown in the machine tool. The air gun on the robot first blows

off the iron filings on the tooling, and then puts the blank into the tooling, forging and pressing, the machine tool starts processing.
When the workpiece processed is finished, the safety door is opened automatically, the robot 2 takes out the workpiece, puts it into the cleaning pool 2, and cleans it. At the same time, the safety door is closed automatically, and the machine tool turret out of the chip cleaning tool, and then opens the safety door automatically.
1PC is sampled when 10PCS are processed, and the robot 2 moves the sampled product to the sampling station 2, and manually returns the water-injection device after the inspection.
When the workpiece is cleaned, the robot 2 puts it into the water spray injection device. After the workpiece is sprayed, it automatically slides to the blanking position, and then manually checks the appearance of the finished product. The operator sequentially puts the finished product into the finished product frame if it's no problem. Now the entire production processing process is Completed.

6. Conclusion

Using Solidworks software to draw the structure of each part and layout of production line, coordinate the placement of four vertical lathes, two robots and conveyor lines within the working radius of the robot. Production line works well through solidworks simulation. The actual automatic production line is set up according to the design program. After a period of trial operation, the automatic production line works well and meets the requirements of enterprise.Provide a solution for the design of mechanical automatic production lines.

7. References

[1] Fu Wang, Dseign and Research of Automated Production Line for Wind Power Brake Housing [D,Nanchang:Nanchang University(2017)
[2] Sheng Yong Lei. Servo-Control System Design of Automatic Production Line Based on PLC and HMI [J]. Applied Mechanics and Materials,457(2012).
[3] Qin Huawei.Research ofAutomatic production line technology forAutomotive hub unit[D]. Wuhan:Huazhong University of Science & Technology(2017)
[4] Zhao Jiayi.Modeling and Simulation of Automative Production Line manufacture core component of Rotary-Vector Reducer[D].Hangzhou:Zhejiang University of Technology(2016)
[5] Zi Hong Zhang, Hong Quan Zhang. The Design of Automatic Control System for Large-ScaleAluminum Ring Rolling Mill [J]. Advanced Materials Research, 443(2012).
[6] Marco A. Jimenez， Susana V.Automation and parameters optimization in production line：a case of study[J].Springer-Verlag London(2012)

Acknowledgement

Supported by Guangdong Youth Innovative Talents Program 2017 (Natural Science) (NO.2017GkQNCX055).

Effect of Magnetic Field on Double Diffusive Natural Convection Inside a Square Cavity with Isothermal Hollow Insert

Satyajit Mojumder[1,a], Sourav Saha[1], Sumon Saha[1,a], Rezwana Rahman[2] and Suvash C. Saha[2]

[1]Department of Mechanical Engineering, Bangladesh University of Engineering and Technology, Dhaka 1000, Bangladesh
[2]School of Mechanical and Mechatronic Engineering, Faculty of Engineering and Information Technology, University of Technology Sydney, NSW, Australia

sumonsaha@me.buet.ac.bd

Abstract. A numerical study is carried out to analyze both heat and mass transfer phenomena in the presence of an external magnetic field inside a square cavity with an isothermal hollow circular insert located at the centre of the cavity. In the present work, the effects of Hartmann number, Lewis number and buoyancy ratio at constant Rayleigh and Prandtl numbers are investigated simultaneously to understand the importance of these parameters on the characteristics of double diffusive natural convection. Galerkin finite element method is employed for the numerical simulations. Grid sensitivity test and code validation are performed prior to confirm the numerical accuracy of the solution. Quantitative comparison is presented by showing the influence of Hartmann and Lewis numbers for different buoyancy ratios on average Nusselt and Sherwood numbers. It is found that increment of Hartmann number results in lower heat and mass transfer rates. Higher value of Lewis number produces elevated mass transfer rate. However, Lewis number has negative impact on heat transfer rate. Moreover, buoyancy ratio has significant effect on heat and mass transfer inside the cavity.

1. Introduction

Obstruction induced convection heat transfer is given prodigious importance because of its recurrence in many real life problems. Heat and mass transfer inside obstructed cavity has a wide range of applications, such as refrigeration systems, solar thermal heaters and coolers, attic space heating and cooling, heat exchangers, cold storages, and so on.

An extensive research has been carried out to analyse heat and mass transfer for different cavities with internal obstacle. Braga and de Lemos [1] studied laminar natural convection in cavities having circular and square rods. They reported that the obstruction had significant influence on heat transfer inside the cavity. House *et al.* [2] investigated the effect of centrally placed square heat conducting body in a vertical enclosure and concluded that heat transfer across the cavity enhanced or reduced depending on the conductivity ratio. Nasrin [3] also studied the effect of centrally placed circular obstacle in a square cavity and showed the impact of Prandtl and Hartmann numbers. Hussain and Hussein [4] investigated natural convection phenomenon for a uniformly heated circular cylinder inside a square cavity. By changing the positions of the circular cylinder, they reported that average Nusselt number behaved nonlinearly as a function of location of the cylinder. Jami *et al.* [5] applied

Lattice Boltzmann method to solve laminar natural convection problem in a square cavity having heat generating and conducting cylindrical body. De and Dalal [6] studied natural convection in a square cavity with horizontally heated square cylinder inside and reported that overall heat transfer changed by changing the aspect ratio of the heat conducting inner cylinder.

Magnetic field is an important parameter that has significant effect on heat and mass transfer for electrically conductive fluid. In fact, the magnetic field creates Lorentz force, which tries to retard the convection current and reduce the heat transfer rate. Magnetic field also has noteworthy effect on mass transfer. Venkatachalappa *et al.* [7] studied effect of magnetic field on heat and mass transfer in a vertical annulus and reported that magnetic field suppressed the double diffusive heat transfer for small buoyancy ratio, whereas for higher buoyancy ratio, magnetic field proved to be effective. They also suggested applying magnetic field perpendicular to the direction of the original flow. More related studies can be found in other literatures [8-12].

In a typical refrigeration system, relatively warmer commodities are placed inside a relatively cooler environment of the chamber. This chamber is surrounded by the heat exchanger tubes through which refrigerants flow and control the temperature of the chamber constantly by evaporation. It is assumed that the surface of the refrigerant carrying tubes are at low temperature under steady-state operation. To model this phenomenon, the present study aims to investigate the effects of magnetic field and buoyancy ratio on a double diffusive natural convection problem inside a square cavity with isothermal hollow insert. Special attention is given to investigate the conjugate effect of Lewis and Hartmann numbers and buoyancy ratio on heat and mass transfer rate.

2. Problem formulation

2.1 Physical modeling

The details of the present model are shown in Figure 1 with appropriate boundary conditions and co-ordinate system. It consists of a square cavity of length L, whose left side wall is heated at temperature T_h and right side wall is cooled at temperature T_c. The top and the bottom walls of the cavity are kept insulated. Concentration of air, with different gas mixture, is approximated to be higher near the left wall (c_h) and lower near the right wall (c_l). The tube is modelled as a hollow cylindrical insert located at the centre of the cavity, which has the constant surface temperature, T_c. An external magnetic field with strength B_o is applied on the left side of the cavity along the positive x-direction. All the boundaries are assumed to be stationary. Viscous heating and radiation effect are neglected for simplifying the problem.

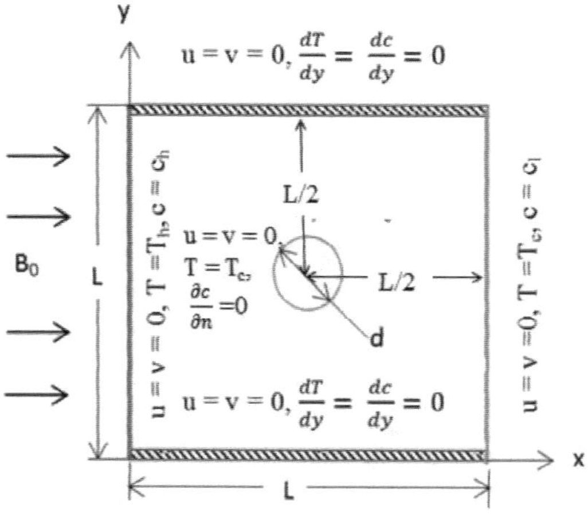

Figure 1. Schematic diagram of the problem with boundary conditions.

2.2 Mathematical modeling

The governing equations for the present problem are based on the conservation laws of mass, linear momentum, concentration and thermal energy in two-dimensions. The flow is assumed to be steady-state and laminar. Constant thermo-physical fluid properties are assumed except density variation in gravitational force term of *y*-momentum equation as related by Boussinesq approximation. Considering the above assumptions, the governing equations in non-dimensional form for the present problem can be written as,

$$\frac{\partial U}{\partial X}+\frac{\partial V}{\partial Y}=0, \tag{1}$$

$$U\frac{\partial U}{\partial X}+V\frac{\partial U}{\partial Y}=-\frac{\partial P}{\partial X}+Pr\left(\frac{\partial^2 U}{\partial X^2}+\frac{\partial^2 U}{\partial Y^2}\right), \tag{2}$$

$$U\frac{\partial V}{\partial X}+V\frac{\partial V}{\partial Y}=-\frac{\partial P}{\partial Y}+Pr\left(\frac{\partial^2 V}{\partial X^2}+\frac{\partial^2 V}{\partial Y^2}\right) +Ra\,Pr\left(\Theta+BrC\right)-Ha^2\,Pr\,V, \tag{3}$$

$$U\frac{\partial \Theta}{\partial X}+V\frac{\partial \Theta}{\partial Y}=\frac{\partial^2 \Theta}{\partial X^2}+\frac{\partial^2 \Theta}{\partial Y^2}, \tag{4}$$

$$U\frac{\partial C}{\partial X}+V\frac{\partial C}{\partial Y}=\frac{1}{Le}\left(\frac{\partial^2 C}{\partial X^2}+\frac{\partial^2 C}{\partial Y^2}\right). \tag{5}$$

The following non-dimensional scales are used to obtain the above non-dimensional equations (1)-(5):

$$\left(X,Y\right)=\frac{\left(x,y\right)}{L},\left(U,V\right)=\frac{\left(u,v\right)}{\alpha/L},P=\frac{pL^2}{\rho\alpha^2}, \quad \Theta=\frac{T-T_c}{T_h-T_c},C=\frac{c-c_l}{c_h-c_l}. \tag{6}$$

In the aforementioned equations, *U* and *V* are the non-dimensional velocities in *X*- and *Y*-directions, *P*, Θ and *C* are the non-dimensional pressure, temperature and concentration, respectively. Dimensionless governing parameters for the present problem are Prandtl number (*Pr*), Rayleigh number (*Ra*), buoyancy ratio (*Br*), Hartmann number (*Ha*) and Lewis number (*Le*). These parameters are defined as,

$$Pr=\frac{\nu}{\alpha},Ra=\frac{g\beta_T\left(T_h-T_c\right)L^3}{\nu\alpha},Br=\frac{\beta_c\left(c_h-c_l\right)}{\beta_T\left(T_h-T_c\right)}, \quad Ha=B_oL\sqrt{\frac{\sigma}{\rho\nu}},Le=\frac{\alpha}{D}. \tag{7}$$

Here, ρ, ν, α and *D* are fluid density, kinematic viscosity, thermal and species diffusivity respectively, β_T and β_c are thermal and compositional expansion coefficient, *g* is gravitational acceleration constant and σ is the electrical conductivity of fluid. The boundary conditions for the present problem in the non-dimensional form are listed in Table 1.

Table 1. Non-dimensional boundary conditions of the problem.

Boundary wall	Velocity	Temperature	Concentration
Left	$U=V=0$	$\Theta=1$	$C=1$
Right	$U=V=0$	$\Theta=0$	$C=0$
Top and bottom	$U=V=0$	$\partial\Theta/\partial Y=0$	$\partial C/\partial Y=0$
Hollow surface	$U=V=0$	$\Theta=0$	$\partial C/\partial N=0$

The performance of the double diffusive natural convection is evaluated by the average Nusselt number (*Nu*) and average Sherwood number (*Sh*) at the left vertical heated wall and those are defined as follows:

$$Nu=-\int_0^1\left(\frac{\partial\Theta}{\partial X}\right)_{X=0}dY,\ Sh=-\int_0^1\left(\frac{\partial C}{\partial X}\right)_{X=0}dY. \tag{8}$$

3. Numerical procedure and validation

Galerkin finite element method is adopted for the numerical solution of the present problem. The entire computational domain is discretised into several triangular mesh elements. The boundary wall is meshed more intensely using finer elements to ensure rapid change of physical variables. The governing differential equations (1) -(5) are transformed into a system of nonlinear algebraic equations and iteration technique is adopted to solve those equations. The convergence criterion set for the present problem is $|\gamma^{m+1} - \gamma^{m}| \leq 10^{-5}$, where γ is the general dependent variable. A grid sensitivity test is also performed to check the numerical accuracy of the present solutions. From the result of grid sensitivity test, it is found that both the average Nusselt and Sherwood numbers at the heated wall are very sensitive to grid size for smaller element number. However, when the element number is more than 4224, both *Nu* and *Sh* become insensitive to the further refinement of mesh elements. Therefore, grid having 4224 elements number is considered as optimum grid for the entire simulation process to reduce the computational effort.

To ensure numerical accuracy of the present model, a validation test has been performed. The present model is validated with the work of Teamah *et al.* [12] on double diffusive natural convection heat transfer in rectangular cavity with the presence of magnetic field. Here the effect of Rayleigh number on the variation of average Nusselt number is observed for $Ha = 10$, $Le = 2$, $Pr = 0.7$, $Br = 1$, aspect ratio of the cavity is 2 and zero inclination angle. Comparison of the model validation test is presented in Figure 2. From the figure, it is evident that the present model is quite reliable to solve a numerical problem with reasonable accuracy.

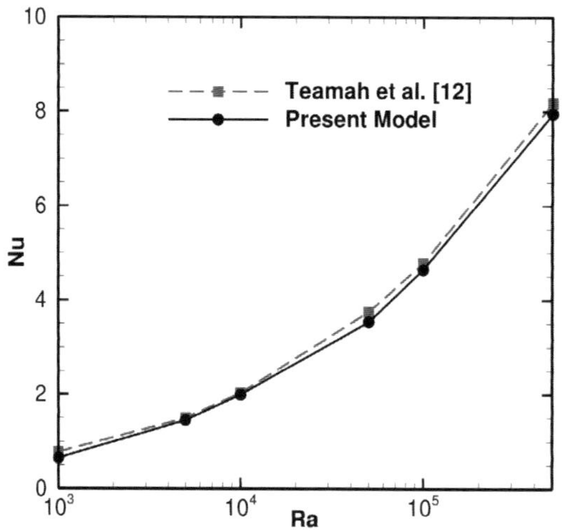

Figure 2. Validation of the present model in terms of average Nusselt number with the results of Teamah et al. [12].

4. Results and discussions

The goal of the present study is to investigate magneto-hydrodynamic heat and mass transfer phenomena inside a square cavity having a cylindrical hollow insert. For the present study, Prandtl number and Rayleigh number are fixed at $Pr = 0.71$ and $Ra = 10^4$ respectively. To investigate mass transfer, buoyancy ratio is varied from $-5 \leq Br \leq 5$ and Lewis number is varied from $1 \leq Le \leq 20$. Hartmann number considered for the present problem is also varied from 0 to 50. The quantitative analysis is done in light of 3D plots incorporating variation of *Nu* and *Sh* against simultaneous variation of *Le* and *Ha*.

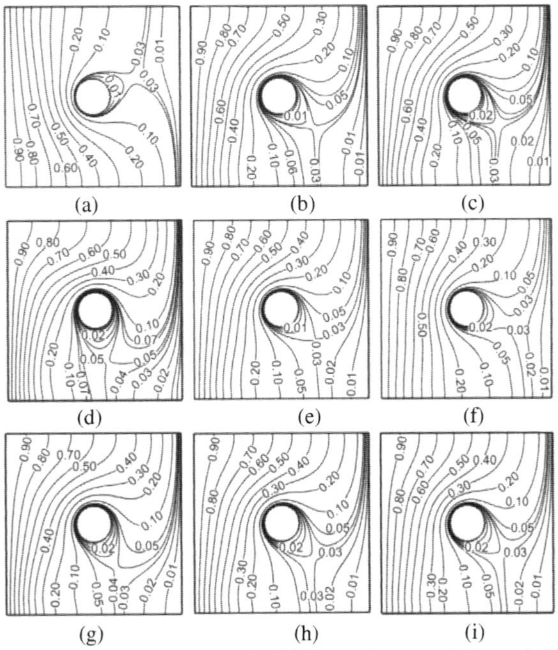

Figure 3. Effects of buoyancy ratio: (a) $Br = -5$, (b) $Br = 1$ and (c) $Br = 5$, Hartmann number: (d) $Ha = 10$, (e) $Ha = 25$ and (f) $Ha = 50$ and Lewis number: (g) $Le = 1$, (h) $Le = 5$ and (i) $Le = 20$ on isotherm contours for $Ra = 10^4$.

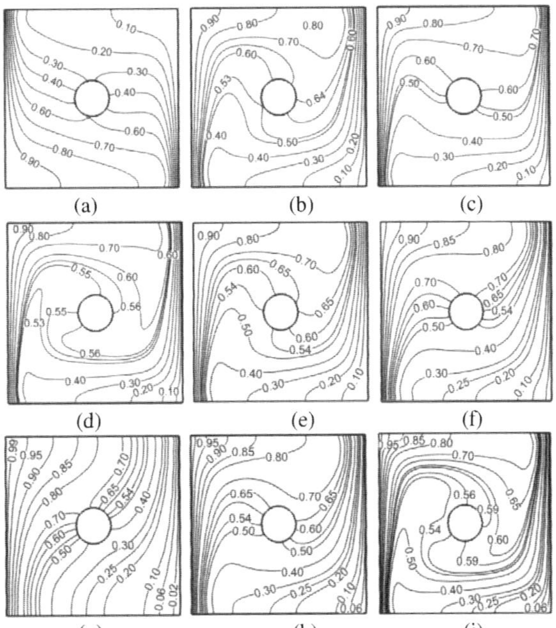

Figure 4. Effects of buoyancy ratio: (a) $Br = -5$, (b) $Br = 1$ and (c) $Br = 5$, Hartmann number: (d) $Ha = 10$, (e) $Ha = 25$ and (f) $Ha = 50$ and Lewis number: (g) $Le = 1$, (h) $Le = 5$ and (i) $Le = 20$ on isoconcentration contours for $Ra = 10^4$.

The effects of buoyancy ratio, Hartmann number and Lewis number on isotherm contours for $Ra = 10^4$ are visualised in Fig. 3(a)-(i) respectively. Isotherm contours for different buoyancy ratios are shown in Fig. 3(a), (b) and (c) at $Le = 10$ and $Ha = 25$. These figures show that for each value of Br, isotherm contours are almost parallel to the left vertical wall and insert wall. The contours are closely packed near the right vertical wall. All of these indicate a steep temperature gradient at those regions which favours conduction over convection. Isotherm contours for different *Ha* are shown in Fig. 3(d), (e) and (f). The contours are nearly parallel near the vertical walls and the cylindrical insert indicating formation of thermal boundary layers in those regions. On the other hand, the isotherm pattern does not vary significantly with the increment of Lewis number as can be seen from Fig. 3(g), (h) and (i).

Figures 4(a), (b) and (c) depict isoconcentration contours for different buoyancy ratios. From figures it can be observed that for $Br = -5$, contours with higher value reside at the bottom of the left wall. Near the top of the left wall and bottom of the right wall, contours are parallel to each other. These indicate strong mass diffusion in those regions. Isoconcentration contours' map for different *Ha* is shown in Fig. 4(d), (e) and (f). Impact of external magnetic field is not that apparent in these figures. However, it is seen that, density of isoconcentration lines becomes less with higher Hartmann number. However, isoconcentration contours undergo drastic changes with varying *Le* (Fig. 4(g), (h) and (i)). With higher values of *Le*, isoconcentration contours are shifting toward the right vertical wall.

A 3D surface plot in Figure 5 shows the combined effect of varying *Le* and *Ha* for different buoyancy ratios on average Nusselt number. From Figure 5, it is found that with the increment of both Lewis and Hartmann numbers, average Nusselt number is reduced. However, those rate and magnitude of reduction vary for different combinations of *Le* and *Ha*. For both $Br = -5$ and 5, with the increment of *Le*, the average Nusselt number decreases at a significant rate, whereas for $Br = 1$, the rate of decrement of average Nusselt number with *Le* is quite less. Again, it is evident that a combination of low Hartmann number and low Lewis number can yield the best possible heat transfer rate. However, impact of Hartmann number is more important on heat transfer rate and even for high *Le*, if low *Ha* can be ensured, quite high heat transfer rate is possible. Moreover, positive and high value of *Br* ensures better heat transfer rate.

Figure 6 shows the combined effect of varying *Le* and *Ha* for different buoyancy ratios ($Br = -5, 1, 5$) on average Sherwood number. From Figure 6, it is observed that the average Sherwood number increases with the increment of the Lewis number and decreases with increasing Hartmann number. A combination of low *Ha* and high *Le* can yield best mass transfer rate at a given value of *Br*. Moreover, change in the value of average Sherwood number is highly influenced by the change of *Le* than that of *Ha*. Hence, if a high value of *Le* is ensured, high *Ha* can also produce satisfactory mass transfer rate. Figure 6 also reveals that positive and high value of *Br* yields higher mass transfer rate.

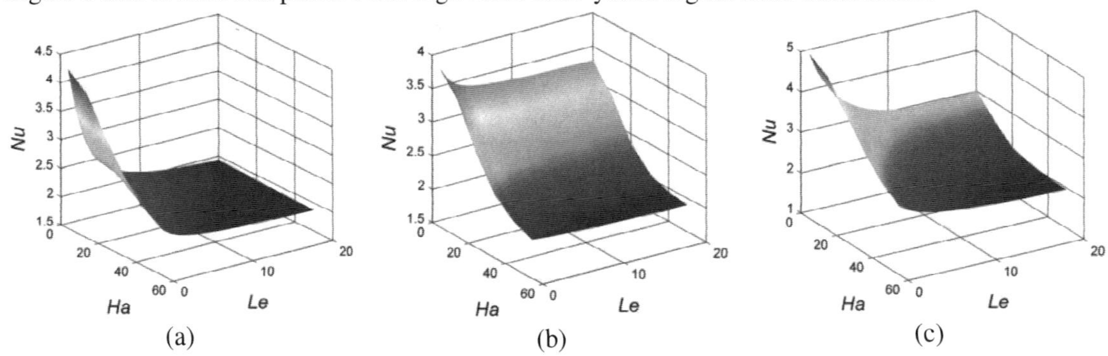

Figure 5. Variation of average Nusselt number as a function of Hartmann and Lewis numbers for different buoyancy ratios: (a) $Br = -5$, (b) $Br = 1$ and (c) $Br = 5$.

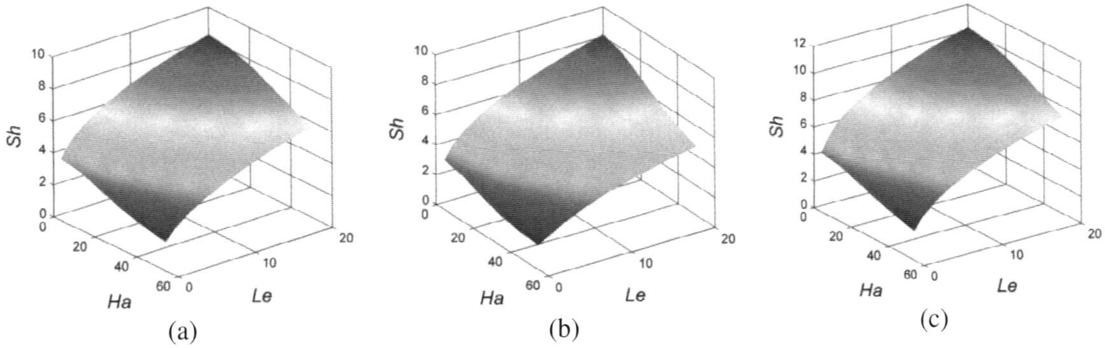

Figure 6. Variation of average Sherwood number as a function of Hartmann and Lewis numbers for different buoyancy ratios: (a) $Br = -5$, (b) $Br = 1$ and (c) $Br = 5$.

5. Conclusions

The present investigation is carried out in order to get a better insight on the influence of Lewis number, Hartmann number and buoyancy ratio on heat and mass transfer in a practical refrigeration model. The following conclusions can be made from the study. Higher values of positive and negative buoyancy ratios ($Br = -5, 5$) cause better heat transfer for low Hartmann number and low Lewis number. When $Br = 1$, the effect of Lewis number is insignificant on heat transfer rate. Increment of buoyancy ratio and Lewis number always assists the mass transfer, whereas increment of Hartmann number reduces the mass transfer. Higher value of Hartmann number retards the flow strength significantly and thus creates hindrance to both heat and mass transfer. With proper combination of governing parameters like low buoyancy ratio ($Br = 1$), low Hartmann number ($Ha = 0 - 10$) and higher Lewis number, it is possible to enhance heat and mass transfer simultaneously. This understanding can be used to design equipment efficiently in a refrigeration system.

6. References

[1] E.J. Braga, M.J.S. de Lemos, Int. Commun. Heat Mass Transf. **32**, 1289–1297 (2005)
[2] J.M. House, C. Beckermann, T.F. Smith, Numer. Heat Transf. A **18**, 213–225 (1990)
[3] R. Nasrin, Int. Commun. Heat Mass Transf. **38**, 625–632 (2011)
[4] S.H. Hussain, A.K. Hussein, Int. Commun. Heat Mass Transf. **37**, 115–1126 (2010)
[5] M. Jami, A. Mezrhab, M. Bouzidi, P. Lallemand, Int. J. Ther. Sci. **46**, 38–47 (2007)
[6] A.K. De, A. Dalal, Int. J. Heat Mass Transf. **49**, 4608–4623 (2006)
[7] M. Venkatachalappa, Y. Do, M. Sankar, Int. J. Eng. Sci. **49**, 262–278 (2011)
[8] M.A. Teamah, Int. J. Therm. Sci. **47**(3), 237–248 (2008)
[9] O. Laguerre, S. Benamara, D. Remy, D. Flick, Int. J. Heat Mass Transf. **52**, 5691–5700 (2009)
[10] X. Zhang, H. Huang, Int. Commun. Heat Mass Transf. **51**, 31–38 (2014)
[11] H. Yamaguchi, X.R. Zhang, X.D. Niu, J. Magn. Magn. Mat. **322**, 698–704 (2010)
[12] M.A. Teamah, A. F. Elsafty, M.Z. Elfeky, E.Z. El-Gazzar, Alex. Engg. J. **50**, 269–282 (2011)

Acknowledgment
The authors gratefully acknowledge the support provided by the Department of Mechanical Engineering, Bangladesh University of Engineering and Technology (BUET) during this research work.

Fractional Sliding Mode Control for Nonlinear Aerospace Systems

Shiqi Zheng[1,2,a], **Bingyun Liang**[1,2]

[1]School of Automation, China University of Geosciences, Wuhan, China
[2]Hubei Key Laboratory of Advanced Control and Intelligent Automation for Complex Systems, Wuhan 430074, China

zhengshiqi1000@foxmail.com

Abstract. This paper focuses on the study of fractional order terminal sliding mode control for nonlinear aerospace systems. Firstly, a novel fractional order integral terminal sliding mode control (FO-I-TSMC) method is proposed for the control of first order nonlinear system. FO-I-TSMC has three attractive advantages: i) Non-singular control law; ii) Elimination of the reaching phase; iii) Calculable finite convergence time. Furthermore, theory analysis is presented to reveal the potential advantages of the FO-I-TSMC method over its integer order counterparts. Secondly, a novel fractional order derivative integral-TSMC (FO-DI-TSMC) method is presented to deal with second order nonlinear system. Finally, FO-DI-TSMC is extended to deal with a general class of higher order control system. Simulation results are given to verify the effectiveness of the proposed methods.

1. Introduction

Aerospace systems, such as satellite, spacecraft, unmanned aerial vehicles, are highly nonlinear systems. High performance aerospace control system should own the features of high tracking response, small overshoot and strong anti-disturbance ability. Conventional PID controller has already been applied in aerospace control system because of its clear functionality, simplicity and ease of implementation [1]. Nevertheless, due to the nonlinearities, unmeasured disturbances and parameters variations, traditional PID controller cannot ensure a sufficiently high control performance. Therefore, various methods of nonlinear control theory, such as predictive control, robust control and fractional order control [2-4], have been proposed for the aerospace system.

Among these advanced control strategies, the sliding mode control is regarded as a powerful method to improve the control performance. The main advantage of sliding mode control is that the control system can exhibit strong robustness properties with respect to both internal parameter variations and external unmeasured disturbances.

The sliding mode control usually consists of two steps: 1) design a sliding surface that ensures the desired error dynamics of the system; 2) drive the system to reach and stay on the specified sliding surface. In conventional sliding mode control, linear sliding surface is adopted to describe the desired performance of the control system. Although the convergence rate of these methods sometimes can be very fast, the system state can only converge to zero in infinite time.

Therefore, in order to further improve the control performance of the closed loop system, nonlinear sliding surfaces are presented to accomplish finite time error convergence. Venkataraman and Gulati

[4] firstly proposed the terminal sliding mode control (TSMC) with a nonlinear sliding surface. Since then, TSMC has attracted a considerable attention both from an academic and industrial point of view. Meanwhile, it has been successfully applied in the servo drive system [5]. TSMC is of interest because the system with finite time convergence can demonstrate very nice characteristics such as higher precision and better robustness. Nevertheless, there exists an intrinsic singular problem in conventional TSMC because of utilizing nonlinear sliding surface. This may result in the instability of the control system.

Fractional calculus is a generalization of the traditional integration and differentiation to the non-integer order [6]. Recently, fractional order controllers have drawn an increasing attention. The extra degrees of freedom from adopting the fractional order made it possible to further improve the control performance compared with integer order controller. Fractional order sliding mode control is one of those fractional order control strategies and it has gained much interest from the control community. In fact, fractional order theory has already been incorporated into the conventional sliding mode control, conventional TSMC and non-singular TSMC methods. For instance, by introducing fractional order dynamic into the traditional sliding surface, BiTao Zhang et al. proposed a novel fractional order sliding mode control strategy for the permanent magnet synchronous motor [7]. Sara Dadras et al. combined the fractional calculus with TSMC to formulate a new fractional order TSMC method for a class of dynamical systems with uncertainty [8]. However, to the best of our knowledge, there is no report on adopting the fractional calculus in the design procedure of I-TSMC and DI-TSMC methods. Therefore, designing fractional order I-TSMC and DI-TSMC methods still remains an open and challenging problem.

Motivated by the previous work, to further improve the tracking response and robustness of the PMLSM servo system, this paper will incorporate the fractional calculus into the design process of I-TSMC and DI-TSMC methods. The detail contributions are as follows:

1) A novel fractional order I-TSMC (FO-I-TSMC) method is proposed for the first order nonlinear system. FO-I-TSMC has three attractive advantages: i) Non-singular control law; ii) Elimination of the reaching phase; iii) Calculable finite convergence time.

2) A novel fractional order DI-TSMC (FO-DI-TSMC) method is presented to deal with second order nonlinear aerospace system.

3) Due to aerospace systems are often high order nonlinear systems. FO-DI-TSMC is further extended to deal with a general class of higher order control system.

2. Preliminaries and definitions

2.1. Fractional calculus

In this section, we will introduce some basic definitions for fractional calculus. Fractional calculus generalizes the ordinary integration and derivation to the non-integer order operators ${}_{t_0}D_t^{\lambda}$. It is given by:

$$ {}_{t_0}D_t^{\lambda} = \begin{cases} \dfrac{d^{\lambda}}{dt^{\lambda}} & if\ \lambda > 0 \\ 1 & if\ \lambda = 0 \\ \int_{t_0}^{t} (dt)^{\lambda} & if\ \lambda < 0 \end{cases} $$

where $\lambda \in \mathrm{R}$ is the fractional order.

There are several definitions for the fractional integration and derivation. The following two definitions will be used in our proposed control method.

Definition 1 [6]. The Riemann-Liouville fractional integration of a function $f(t)$ is defined as:

$$ {}_{t_0}D_t^{\lambda} f(t) = \frac{1}{\Gamma(-\lambda)} \int_{t_0}^{t} (t-\tau)^{-\lambda-1} f(\tau) d\tau $$

The fractional derivation of is defined as

$$ {}_{t_0}D_t^{\lambda} f(t) = \frac{d^m}{dt^m}\left[\frac{1}{\Gamma(m-\lambda)}\int_{t_0}^{t}\frac{f(\tau)}{(t-\tau)^{\lambda-m+1}}d\tau\right] $$

Definition 2 [6]. The Grunwald-Letnikov fractional integration and derivation of a function $f(t)$ is defined as:

$$ {}_{t_0}^{L}D_t^{\lambda} f(t) = \lim_{h\to 0}\frac{1}{h^{\lambda}}\sum_{i=0}^{[(t-t_0)/h]}(-1)^i\binom{\lambda}{i}f(t-ih) $$

where h represents the sampling time, $[(t\text{-}t_0)/h]$ is the rounding function and

$$ \binom{\lambda}{i} = \frac{\lambda(\lambda-1)\cdots(\lambda-i+1)}{i!} $$

Remark 1. Note that for the number of addends in the above equation becomes enormously large. This cannot be implemented in practice. Using the short memory principle in Ref. [6], the Grunwald-Letnikov definition can be approximated as

$$ {}_{t_0}^{L}D_t^{\lambda} f(t) \approx {}_{t-L}^{L}D_t^{\lambda} f(t) $$

where L is the memory length.

2.2. Control objectives

We will consider the control problem for the following three classes of nonlinear systems, i.e., first order, second order and high order nonlinear systems. These nonlinear systems can describe a wide class of aerospace nonlinear systems, such as satellite, spacecraft, unmanned aerial vehicles, servo drive systems etc.

1) First order nonlinear systems

$$ \dot{e}_v(t) = f_v(e_v) + g_v(e_v)i_q + d(t) $$

where e_v is the velocity tracking error, f and g are nonlinear systems, d is disturbance.

2) Second order nonlinear systems

$$ \ddot{e}_\theta(t) = f_\theta(\boldsymbol{e}_\theta) + g_\theta(\boldsymbol{e}_\theta)i_q + d(t) $$

where e is the position tracking error, f and g are nonlinear systems, d is disturbance.

3) High order nonlinear systems

$$ x^{(n)}(t) = \bar{f}(\boldsymbol{x}) + \bar{g}(\boldsymbol{x})u + \bar{d}(t) $$

where x represents the state, f and g are nonlinear systems, d is disturbance.

Considering the tracking reference, he above equation can be written as

$$ \begin{aligned} e^{(n)}(t) &= x_r^{(n)} - \bar{f}(\boldsymbol{x}_r - \boldsymbol{e}) - \bar{g}(\boldsymbol{x}_r - \boldsymbol{e})u - \bar{d}(t) \\ &= f(\boldsymbol{e}) + g(\boldsymbol{e})u + d(t) \end{aligned} $$

where e is the tracking error.

The control objective is to design a controller u such that the tracking error is minimized.

3. Fractional order integral terminal sliding mode control

We will consider the sliding mode control of first order system. The sliding mode control for the first order system can be adopted for the speed control of some aerospace system. Next, we will present the FO-I-TSMC method. FO-I-TSMC consists of two different types, sign and fractional integral sliding mode.

3.1. Type I: fractional-order sign integral sliding mode

The fractional-order sign integral sliding surface is designed as:

$$ s(t) = e(t) + \alpha e_I(t) $$

$$ \dot{e}_I(t) = P(e(t)) = \begin{cases} {}_0D_t^{1-\lambda}[\mathrm{sgn}(e(t))] & if\ \ e(t)\neq 0 \\ 0 & if\ \ e(t)=0 \end{cases} $$

where $\alpha > 0$, $0 < \lambda < 2$ and we also have

$$ e_I(0) = -e(0)/\alpha . $$

By adopting the above definition, we have the following results:

Theorem 1. *The first order system, i.e. the system described by (1) is in the sliding surface all the time, i.e.* $s(t) \equiv 0, \forall t \geq 0$, *if the following controller is applied*

$$u(t) = g(e)^{-1}\{-f(e) - \alpha P(e(t)) - \eta \operatorname{sgn}(s(t))\}$$

where $\eta > \delta \geq |d(t)|$.

When using the above control action, the convergence time of the system will be finite.

Theorem 2. *Suppose the above control law is adopted, the tracking error e will converge to zero in finite time* $T_s = (\frac{|e(0)|\Gamma(\lambda+1)}{\alpha})^{1/\lambda}$.

3.2. Type II: fractional-order fractional integral sliding mode

The fractional-order sign integral sliding surface is designed as:

$$s(t) = e(t) + \alpha e_I(t)$$

$$\dot{e}_I(t) = Q(e(t)) = \begin{cases} {}_0D_t^{1-\lambda}[(e(t)^{q/p}] & if \ \ e(t) \neq 0 \\ 0 & if \ \ e(t) = 0 \end{cases}$$

where $\alpha > 0$, $0 < \lambda < 2$ and we also have

$$e_I(0) = -e(0)/\alpha .$$

By adopting the above definition, we have the following results:

Theorem 3. *The first order system, i.e. the system described by (1) is in the sliding surface all the time, i.e.* $s(t) \equiv 0, \forall t \geq 0$, *if the following controller is applied*

$$u(t) = g(e)^{-1}\{-f(e) - \alpha Q(e(t)) - \eta \operatorname{sgn}(s(t))\}$$

where $\eta > \delta \geq |d(t)|$.

The convergence time of the system will also be finite.

Theorem 2. *Suppose the above control law is adopted, the tracking error e will converge to zero in finite time* $T_s = (\frac{|e(0)|^{1-q/p}\Gamma(\lambda+1)}{\alpha(1-q/p)})^{1/\lambda}$.

Remark 4. The proposed FO-I-TSMC methods have some nice features which are similar to the IO-I-TSMC [12]. 1) According to Theorems 1 and 3, it can be concluded that there is no reaching phase and the control system is always kept in the sliding surface. This indicates that the servo system could have better robustness to parameter uncertainties and faster response. 2) The tracking error can converge to zero in finite time. Moreover, the convergence time can be calculated easily based on Theorems 2 and 4. 3) We can see that the FO-I-TSMC methods do not have singular problems compared with the conventional TSMC. 4) The gain only depends on the bound of the uncertainty. Therefore, the gain can be adjusted by using some adaptation mechanism such as neural networks or gradient descent methods.

4. Fractional order derivative-integral terminal sliding mode control

As can be seen, the results in Section 3 are only suitable for the first order system. Also the speed loop of the aerospace nonlinear system is typically a first order system. For the control of position loop or other higher order system, the FO-I-TSMC cannot be applied. Therefore, in this section the fractional order derivative-integral TSMC method (FO-DI-TSMC) is proposed to deal with the output tracking problem for higher order system. First, we will consider the control of second order system. Next, we will consider a general class of high order system.

4.1. FO-DI-TSMC for a class of second order system

FO-DI-TSMC for second order system is in fact a combination of the non-singular fractional order TSMC in Refs. [10-11] and our proposed FO-I-TSMC.

The fractional-order sign integral sliding surface is defined as:

$$e_{D_1}(t)=[{}_0D_t^{\gamma}\ddot{e}_{D_0}(t)]^{p_1/q_1}+\beta_1 e_{D_0}(t)$$
$$s(t)=e_{D_1}(t)+\alpha e_I(t)$$

where

$$e_I(0)=-e_{D_1}(0)/\alpha\,.$$

According to the above definition, the sliding surface can be expressed as:

$$\dot{s}(t)=\frac{p_1}{q_1}[{}_0D_t^{\gamma}\ddot{e}_{D_0}(t)]^{(p_1/q_1-1)}\cdot{}_0D_t^{\gamma+1}\ddot{e}_{D_0}(t)+\psi$$
$$=\frac{p_1}{q_1}[{}_0D_t^{\gamma}\ddot{e}_{D_0}(t)]^{(p_1/q_1-1)}\cdot{}_0D_t^{\gamma+1}[f(\boldsymbol{e})+g(\boldsymbol{e})u+d(t)]+\psi$$

where $\psi=\beta_1\dot{e}_{D_0}+\alpha\dot{e}_I$

Let

$$F_\pi\equiv\begin{cases}[{}_0D_t^{\gamma}\ddot{e}_{D_0}(t)]^{(p_1/q_1-1)} & if\ |[{}_0D_t^{\gamma}\ddot{e}_{D_0}(t)]^{(p_1/q_1-1)}|\geq\varepsilon\\ \varepsilon & if\ |[{}_0D_t^{\gamma}\ddot{e}_{D_0}(t)]^{(p_1/q_1-1)}|<\varepsilon\end{cases}$$

Then we have

$$|(F_\pi-[{}_0D_t^{\gamma}\ddot{e}_{D_0}(t)]^{(p_1/q_1-1)})F_\pi^{-1}|\leq\varphi<1$$

Using the above result, we have

Theorem 5. The overall system is in the sliding mode by Definition 5 all the time, *i.e.* $s(t)\equiv 0, \forall t\geq 0$, *if the following controller is applied*

$$u(t)=g(\boldsymbol{e})^{-1}\{-f(\boldsymbol{e})$$
$$-{}_0D_t^{-\gamma-1}[K_e|\psi|(\frac{q_1}{p_1}F_\pi^{-1})\operatorname{sgn}(s(t))+\eta\operatorname{sgn}(s(t))]\}$$

Theorem 6. *Suppose the above control law is adopted, and* $\dot{e}_I(t)$ *is designed as* $P(e(t))$ *the tracking error e will converge to zero in finite time*

$$\begin{cases}T_s=(\frac{|e(0)|\Gamma(\lambda+1)}{\alpha})^{1/\lambda}+\frac{|e(t_1)|^{1-q_1/p_1}}{\beta_1(1-q_1/p_1)} & if\ \ \gamma=-1\\ T_s\leq(\frac{|e(0)|\Gamma(\lambda+1)}{\alpha})^{1/\lambda}+\frac{2|e(t_1)|}{|\dot{e}(t_1)|} & if\ \ 0<\gamma<1\end{cases}.$$

Theorem 7. *Suppose the above control law is adopted, and* $\dot{e}_I(t)$ *is designed as* $Q(e(t))$ *the tracking error e will converge to zero in finite time*

$$\begin{cases}T_s=(\frac{|e(0)|^{1-q/p}\Gamma(\lambda+1)}{\alpha(1-q/p)})^{1/\lambda}+\frac{|e(t_1)|^{1-q_1/p_1}}{\beta_1(1-q_1/p_1)} & if\ \ \gamma=-1\\ T_s\leq(\frac{|e(0)|^{1-q/p}\Gamma(\lambda+1)}{\alpha(1-q/p)})^{1/\lambda}+\frac{2|e(t_1)|}{|\dot{e}(t_1)|} & if\ \ 0<\gamma<1\end{cases}$$

5. FO-DI-TSMC for a class of high order system

In this subsection, we will consider the control of a general high order system described by (3). To begin with, we will introduce the fractional order derivative-integral sliding surface.

The fractional-order sign integral sliding surface is defined as:

$$e_{D_1}(t)=[\dot{e}_{D_0}(t)]^{p_1/q_1}+\beta_1 e_{D_0}(t)$$
$$e_{D_2}(t)=[\dot{e}_{D_1}(t)]^{p_2/q_2}+\beta_2 e_{D_1}(t)$$
$$\cdots$$
$$e_{D_{n-1}}(t)=[\dot{e}_{D_{n-2}}(t)]^{p_{n-1}/q_{n-1}}+\beta_{n-1}e_{D_{n-2}}(t)$$
$$s(t)=e_{D_{n-1}}(t)+\alpha e_I(t)$$
$$e_I(0)=-e_{D_{n-1}}(0)/\alpha$$

From the above definition, we can derive

$$\begin{aligned}\dot{s}(t) &= \dot{e}_{D_2} + \alpha \dot{e}_I \\ &= \frac{p_2}{q_2}[\dot{e}_{D_1}]^{(p_2/q_2-1)}\ddot{e}_{D_1} + \beta_2 \dot{e}_{D_1} + \alpha \dot{e}_I \\ &= \frac{p_2}{q_2}[\dot{e}_{D_1}]^{(p_2/q_2-1)}\{\frac{p_1}{q_1}[\dot{e}_{D_0}]^{(p_1/q_1-1)}\ddot{e}_{D_0} + \beta_1 \dot{e}_{D_0}\}' + \beta_2 \dot{e}_{D_1} + \alpha \dot{e}_I \\ &= \frac{p_2}{q_2}[\dot{e}_{D_1}]^{(p_2/q_2-1)}\{\frac{p_1}{q_1}[\dot{e}_{D_0}]^{(p_1/q_1-1)}\dddot{e}_{D_0} + \frac{p_1}{q_1}(\frac{p_1}{q_1}-1)[\dot{e}_{D_0}]^{(p_1/q_1-2)}(\ddot{e}_{D_0})^2 + \beta_1 \ddot{e}_{D_0}\} \\ &\quad + \beta_2 \dot{e}_{D_1} + \alpha \dot{e}_I\end{aligned}$$

$$\begin{aligned}\phi &= \frac{p_2}{q_2}[\dot{e}_{D_1}]^{(p_2/q_2-1)}\{\frac{p_1}{q_1}(\frac{p_1}{q_1}-1)[\dot{e}_{D_0}]^{(p_1/q_1-2)}(\ddot{e}_{D_0})^2 + \beta_1 \ddot{e}_{D_0}\} \\ &= \frac{p_2}{q_2}\{\frac{p_1}{q_1}[\dot{e}_{D_0}]^{(p_1/q_1-1)}\ddot{e}_{D_0} + \beta_1 \dot{e}_{D_0}\}^{(p_2/q_2-1)}\{\frac{p_1}{q_1}(\frac{p_1}{q_1}-1)[\dot{e}_{D_0}]^{(p_1/q_1-2)}(\ddot{e}_{D_0})^2 + \beta_1 \ddot{e}_{D_0}\}\end{aligned}$$

Then, we have the following results

Theorem 8. The overall system is in the sliding mode by Definition 6 all the time, *i.e.* $s(t) \equiv 0, \forall t \geq 0$, *if the following controller is applied*

$$\begin{aligned}u(t) &= g(\mathbf{e})^{-1}\{-f(\mathbf{e}) \\ &\quad - K_e |\psi| (\prod_{i=1}^{n-1} \frac{q_i}{p_i} F_\pi^{-1}) \operatorname{sgn}(s(t)) - \eta \operatorname{sgn}(s(t))\}\end{aligned}.$$

Meanwhile, the convergence time is given by

Theorem 9. Suppose the above control law is adopted and $\dot{e}_I(t)$ is designed as $P(e_{D_{n-1}}(t))$, then the tracking error $e(t)$ will converge to zero in finite time

$$T_s = (\frac{|e(0)|\Gamma(\lambda+1)}{\alpha})^{1/\lambda} + \sum_{i=1}^{n-1} \frac{|e_{D_{(i-1)}}(t_i)|^{1-q_i/p_i}}{\beta_i(1-q_i/p_i)}.$$

Theorem 10. Suppose the above control law is adopted and $\dot{e}_I(t)$ is designed as $Q(e_{D_{n-1}}(t))$, then the tracking error $e(t)$ will converge to zero in finite time

$$T_s = (\frac{|e(0)|^{1-q/p}\Gamma(\lambda+1)}{\alpha(1-q/p)})^{1/\lambda} + \sum_{i=1}^{n-1} \frac{|e_{D_{(i-1)}}(t_i)|^{1-q_i/p_i}}{\beta_i(1-q_i/p_i)}.$$

6 Simulation

In this section, simulations are conducted to verify the effectiveness of the proposed FO-I-TSMC and FO-DI-TSMC methods. The proposed methods are mainly adopted to control the servo motor system, which has been widely applied in aerospace engineering. The simulation environment is the MATLAB 2014a. The model of the motor is identified from a real experimental platform shown in Fig. 1.

First, we will focus on the speed control of the PMLSM system. A step speed reference is exerted on the servo system. Type I IO-I-TSMC and FO-I-TSMC are adopted. The control parameters are set as: $\alpha = 1, \eta = 1$. The tracking responses are shown in Fig. 2. It can be seen that the rising time of FO-I-TSMC with $0 < \lambda < 1$ is the shortest. FO-I-TSMC with $\lambda > 1$ has a sluggish tracking response. Therefore, it can be concluded that smaller λ will result in faster tracking. Meanwhile, it can also be seen from the partial enlarged view in Fig. 1 that in the steady state FO-I-TSMC with $\lambda > 1$ has the smallest chattering. However, FO-I-TSMC with $0 < \lambda < 1$ has the largest. Hence, we can conclude that a larger λ can reduce the chattering of the system.

Fig. 3 also shows the variation of $P(e(t))$. It can be seen that for IO-I-TSMC $|P(e(t))|$ is always equal to one. Nevertheless, for FO-I-TSMC, $|P(e(t))| > 1$ for $0 < \lambda < 1$ which indicates a faster tracking, and $|P(e(t))| < 1$ for $\lambda > 1$ which implies a smaller overshoot and chattering.

In order to obtain a faster tracking response and smaller chattering, one can use a time varying fractional order I-TSMC. A larger λ may be needed for the tracking process and a smaller λ for the steady state. We select $\lambda=0.8$ when $|e(t)|\geq 0.01$ and $\lambda=1.4$ when $|e(t)|<0.01$. The tracking response of the time varying fractional order I-TSMC is illustrated in Fig. 4. It can be seen that compared with the IO-I-TSMC, it not only has faster tracking response but also a smaller chattering.

Figure.1. Platform of servo system.

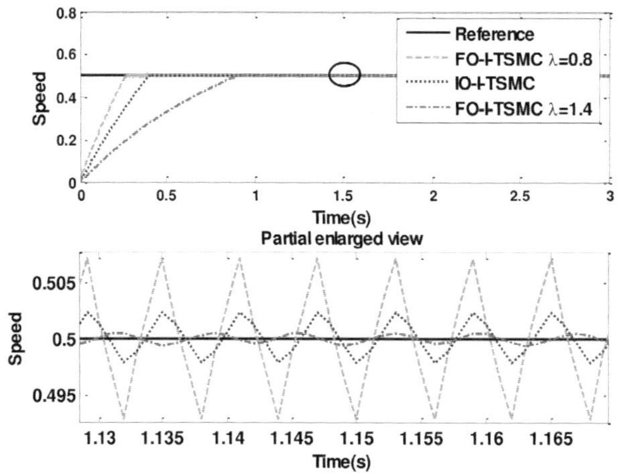

Figure. 2. Step responses for Type I I-TSMC.

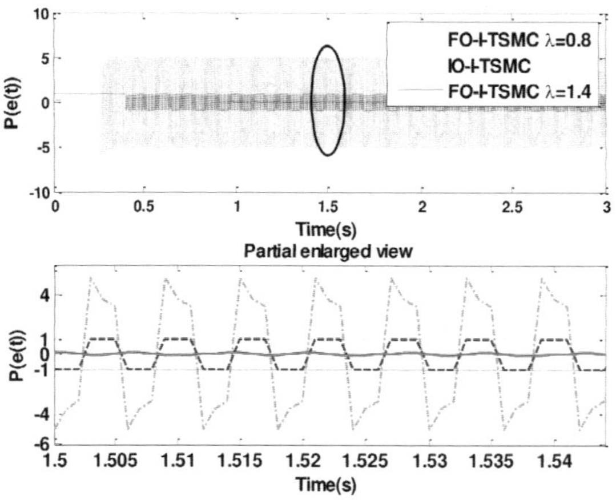

Figure.3. Variation of $P(e(t))$ for step response.

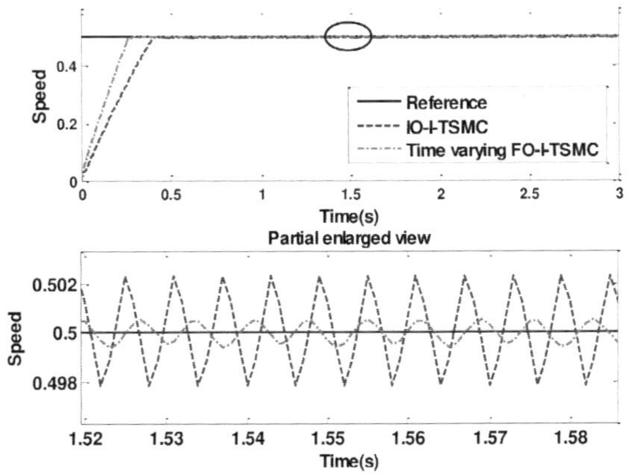

Figure. 4. Step responses for time varying Type I I-TSMC.

7. Conclusions

In this paper, we propose two kinds of fractional order terminal sliding mode control methods, i.e., FO-I-TSMC and FO-DI-TSMC, to improve the control performance of some aerospace nonlinear system. FO-I-TSMC can be applied in the speed loop of the servo system. Theory analysis and simulation results all verify the effectiveness of FO-I-TSMC over its integer order counterparts. FO-DI-TSMC can be applied in the position loop of the servo system.

8. References

[1] Lu S., S. Zheng, X. Tang, B. Song, J. Process Contr. Vol. 23, No. 10, pp. 1455-1464 (2013).
[2] Zheng S., X. Tang, B. Song, P. I. Mech. Eng. C-J. Mec. Vol. 228, No. 17, pp. 3218-3239 (2013).
[3] Yin C., Y.Q. Chen, S. Zhong, Automatica. Vol. 50, No. 12, pp. 3173-3181 (2014).
[4] Venkataraman S.T., S. Gulati, J. Dyn. Syst.-T. ASME. Vol. 115, No. 3, pp. 554-560 (1993).
[5] Qi L., H. Shi, Neurocomputing. Vol. 115, pp. 23-30 (2013).
[6] Rhouma A., F. Bouani, IET Control Theory A. Vol. 8. No. 17, pp. 1986-1994 (2014).
[7] Zhang B., Y. Pi, Y. Luo, "Fractional order sliding-mode control based on parameters auto-tuning for velocity control of permanent magnet synchronous motor," ISA T. Vol. 51, No. 5, pp. 649-659 (2012).
[8] Dadras S., H. Momeni, Commun. Nonlinear. Sci. Vol. 17, No. 1, pp. 367-377 (2012).
[9] Chiu C.-S., Automatica. Vol. 48, No. 2, pp. 316-326 (2012).
[10] Dadras S., H. Momeni, Commun. Nonlinear. Sci. Vol. 17, No. 1, pp. 367-377 (2012).
[11] Zhao G., Math. Probl. Eng. pp. 1-10 (2013).
[12] Yin C., Y.Q. Chen, S. Zhong, Automatica. Vol. 50, No. 12, pp. 3173-3181 (2014)

Acknowledgement

This work is supported by National Natural Science Foundation of China (61703376), the State Scholarship Fund from China Scholarship Council (201806415019)

Kinematics Test and Evaluation of Tool Axis Direction Error for Five-axis Machine Tools

L Zhong[1], F Ren[2], L Guo[2], Q Bi[1] and Y Wang[1]

[1] State Key Laboratory of Mechanical System and Vibration, School of Mechanical Engineering, Shanghai Jiao Tong University, Shanghai, 200240, PR China
[2] Shanghai Aerospace Equipments Manufacture, Shanghai, 200245, PR China

zhonglei@sjtu.edu.cn

Abstract. This paper presents a kinematics detection and evaluation method for tool axis direction error of five-axis machine tools. By measuring the trajectory errors of two tool center positions corresponding to two different tool lengths, the tool axis direction error can be obtained by combining the two measurements. Compared with the evaluation method in ISO10791.6.2014 which can only evaluate the tool center position errors corresponding to a single tool length, the proposed method can infer the processing accuracy corresponding to different tool lengths and to evaluate the comprehensive accuracy of five-axis flank milling. The proposed method is verified through simulations and experiments on a five-axis gantry machine tool.

1. Introduction

Five-axis machine tool is an effective tool for processing complex feature parts. Its dynamic accuracy determines the quality and machining efficiency of the workpiece. Kinematics test is an effective method which can evaluate the dynamic performance of five-axis machine tools. Many scholars have designed a variety of test paths to test the performance of five-axis machine tools by using a variety of test instruments, such as double ball bar (DBB) [1], R-test [2], laser tracking interferometer [3] and so on. Masaomi et al. [1] designed circular trajectory measurement methods using DBB to evaluate the accuracy of the rotary axes. The authors also present the evaluation method of spherical deviation error by using DBB [4] and the S trajectory deviation test method by using R-test [5]. Ibaraki et al. [6] constructed an error map of rotary axes by R-test. In ISO10791-6:2014 [7], four kinematics test methods for dynamic accuracy evaluation of five-axis machine tools with three typical structures are defined. Among them, AK3 is a non-cutting method equivalent to cone-frustum milling by using DBB and AK4 checks the deviations of the tool center point trajectory, which is ideally a fixed point in the workpiece coordinate system (WCS).

The above-mentioned kinematics test method can effectively detect the dynamic error of the tool center point (TCP), but it cannot evaluate the dynamic error of the tool axis direction (TAD). The envelope surface formed by tool movement in five-axis machining process determines the dimensional accuracy of the parts. The position of the tool is determined by both the tool center point and the tool axis direction. When different tool lengths or five-axis flank milling are used, it is impossible to evaluate the machining accuracy according to the TCP error of a point. In this paper, a kinematics testing and evaluation method for tool axis direction error of five-axis machine tools is presented. By measuring and synthetically analyzing the trajectory errors of two tool center positions corresponding to two different tool lengths, the TAD error and TCP error can be obtained by combining the two measurements. With the data of tool axis direction error, the performance of five-axis machine tools can be evaluated more comprehensively.

2. Evaluation Method of Tool Axis Direction Error

The principle of comprehensive evaluation index of TCP and TAD error proposed in this paper is as shown in figure 1(a). P_0 is the initial tool center point and P_t is the point on the tool axis. The distance between them is L and the length of the tool is D. The machine tool makes the tool reach theoretical position P_n and P_{tn} according to the test path. In fact, because of machine tool error, the actual reaching position of tool center point is P_r, and the point on tool axis reaches P_{tr}.

The distance e_p between P_n and P_r is the TCP error E_{TCP}, which is the evaluation index of the TCP error of five-axis machine tool stipulated in ISO10791.6-2014. e_t is the distance between P_{tn} and P_{tr} . e_t and e_p is different due to the TAD error. So the tool axis direction error E_{TAD} can be defined as the ratio of the $(e_t - e_p)$ to L. Therefore, the evaluation index of tool tip and tool axis direction error can be determined by equation (1) and equation (2).

$$E_{TCP} = \max\left(\left|\overrightarrow{e_{pi}}\right|\right) \quad (1)$$

$$E_{TAD} = \max\left(\frac{\left|\overrightarrow{e_{ti}}-\overrightarrow{e_{pi}}\right|}{L}\right) \quad (2)$$

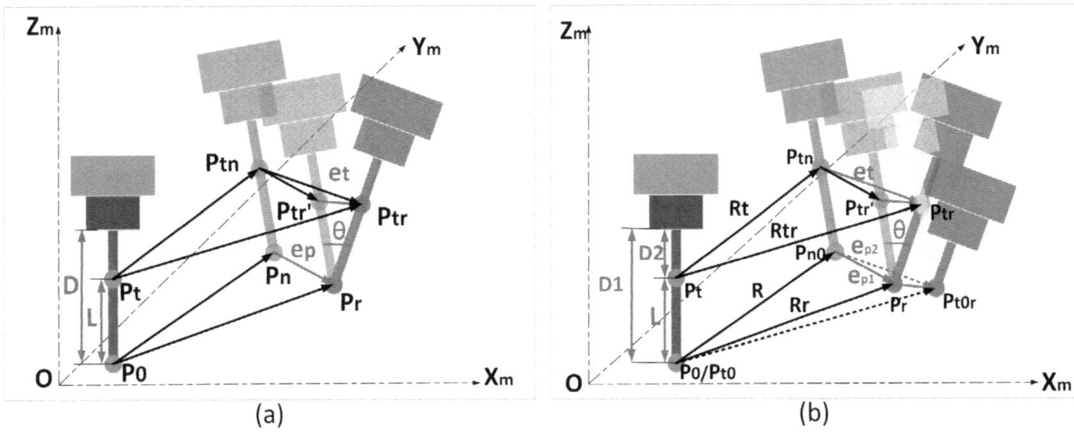

Figure 1. Geometric relationship between TCP and TAD error

3. Measuring method of tool axis direction error

This section uses the AK3 and AK4 test paths in ISO standards to detect the TCP and TAD error.

As shown in figure 2, the proposed method is to measure AK3 motion twice with a DBB. The same measuring NC code is used for the two measurements, but different tool lengths D1 and D2 are used for two actual measurements, and different tool lengths D1 and D2 are set in the CNC system. Because of the different tool lengths, the positions of the three linear axes corresponding to the two measurements are different, but the motion of the two rotary axes is the same.

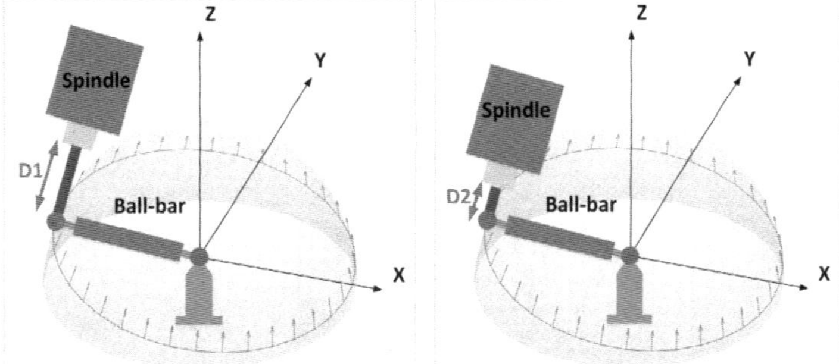

Figure 2. The measuring path of AK3 test with different tool length

As shown in figure 1(b), in the second measurement with tool length D2, because the processing program is unchanged, that is, the theoretical trajectory of P_t point is the same as that of P_0 point, the P_t point in the direction of the original tool axis is moved to P_0 point through the interpolation motion of the three axes of X\Y\Z. Assuming that the additional errors caused by the change of linear axes travel are ignored, the errors measured twice can be considered to be unchanged. When the machine tool moves to P_{tn} point, the actual position is still P_{tr}. The linear axis interpolates in the direction of the theoretical tool axis and moves the P_{tr} point to P_{t0r} point. From the geometric relationship in figure 1(b), it can be seen that the TCP P_{t0r} error obtained by the second measurement is equal to the point P_{tr} error in the direction of the tool axis. Therefore, the TAD error can be calculated approximately by the equation (3).

$$E_{TAD} \approx \max\left(\frac{|\overrightarrow{e_{p2i}} - \overrightarrow{e_{p1i}}|}{L}\right) \tag{3}$$

Because the DBB can only measure one-dimensional distance information, it is impossible to calculate directly by equation (3), but the actual distance R_r and R_{tr} can be measured by the DBB. So equation (4) and equation (5) can be used to calculate the TCP and TAD error.

$$E_{TCP-AK3} = \max(|R_{ri} - R_i|) \tag{4}$$

$$E_{TAD-AK3} = \left|\frac{\max|((R_{tri} - R_{ti}) - (R_{ri} - R_i))|}{D_1 - D_2}\right| \tag{5}$$

It is worth mentioning that, because the tool length L is generally less than 100 mm, the additional machine tool error caused by the change of linear axis 200 mm stroke is generally less than 0.004 mm, which can be neglected compared with the roundness 0.08 mm required by the ISO10791.7.2014 [8] for the milling of cone frustum.

As mentioned in [4], R-test can measure the TCP error in three directions at one time, so it is more convenient to use R-test to perform AK4 measurement. Its basic principle and error calculation method are described in detail in [4]. The AK4 measurement method of TAD error is similar to AK3. Two AK4 measurements are made by R-test. The TAD and TCP error can be obtained by equation (1) and (2).

4. Simulation Analysis of Tool Axis Direction Error

The machine tool structure adopted in the simulation and experiment in this paper is a five-axis machine tool with two rotary axes in the spindle head and the kinematics model is described in [4]. The flowchart of simulation strategy is shown in Figure 3. The tool location coordinates (x, y, z, i, j, k) of the test path in the WCS are post-processed to obtain the axis motion (X, Y, Z, A, C) in the machine tool coordinate system. The second-order dynamic control model is used to simulate the control of the servo system on each axis. $\omega_n = 100Hz$ is the natural frequency and ζ =0.7 is the damping ratio. The proportional gain of the position control loop is K. After the servo system simulation, the actual positions of each axis (X', Y', Z', A', C') are obtained, and then the position values are transformed into actual tool locations in WCS by the forward kinematic model of the machine tool with the volumetric errors. After processing, the actual tool position is changed from P (x, y, z) to P' (x', y', z'), as well as the tool axis directions U (x, y, z) to U' (x', y', z'). Finally, the simulation results of AK 3 and AK4 are obtained by using the calculation method described in Section 3.

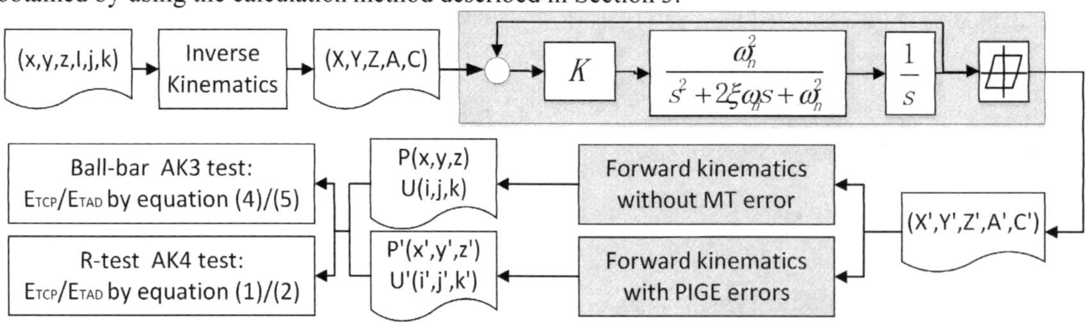

Figure 3. Flowchart of simulation strategy

4.1. Influence of geometric errors

Geometric accuracy is the basis of dynamic motion accuracy of five-axis machine tools. The 8 PIGEs associated with rotary axes are set either 0.01mm for linear errors or 0.01° for angular errors. The tool length is set to 100mm. Then a simulated contour error trajectory in the WCS can be obtained as shown in Figure4(a), the red lines are the tool axes with errors, and the blue lines are the theoretical tool axes. For easy observation, the deviation is enlarged by 100 times. The TAD and TCP errors results are shown in Table 1. Figure4 (b) show the error maps of the tool center point and the point on tool axis, which are shown in red and blue respectively. It can be seen that in AK3 and AK4 tests, the two points on tool axis have different error trajectories, which is mainly caused by the TAD error.

4.2. Influence of proportional gain mismatch

The proportional gain K of A axis is set to 35 (1/s), and the other axis is 40(1/s), respectively. This can be used to simulate the error caused by the slow servo response of A axis. From Table 1 and Figure 4 (c), it can be seen that the servo mismatch of rotary axis will also produce a large TAD error. In addition, because the motion range and acceleration and deceleration requirements of AK4 test are larger than those of AK3 test, the TAD error in AK4 test caused by slow servo response of A axis is larger.

4.3. Influence of positioning error

It is assumed that the positioning error of A axis has a positive linear trend with a slope of 0.001°, i.e. the positioning error is 0.09° in the range of 90°. From Table 1 and Figure 4(d), it can be seen that the positioning error of A axis will directly produce the equivalent TAD deviation because the swing of A axis directly affects the direction of tool axis.

From the above simulation results, it can be seen that many error factors of rotary axis will lead to the TAD and TCP error. The two-point measurement method can evaluate the TAD error caused by the angular error, so as to better reflect the comprehensive performance of the machine tool.

Table 1. TAD error simulated with different error sources

	PIGEs		gain mismatch		positioning error	
Error term	AK3 test	AK 4test	AK3 test	AK 4test	AK3 test	AK 4test
First TCP error	55.5 μm	46.2 μm	40.8 μm	94.1 μm	61.9 μm	157.0 μm
Second TCP error	23.1 μm	24.5 μm	5.5 μm	2.3 μm	11.0 μm	15.0 μm
TAD error	81.9 sec	81.9 sec	115.0 sec	195.2sec	107.7 sec	323.8 sec

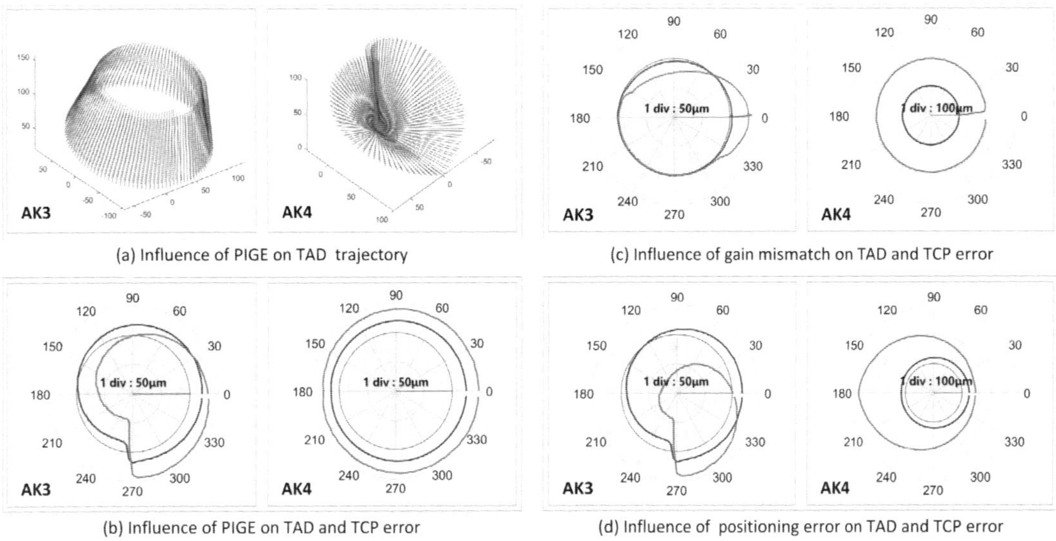

Figure 4. Influence of different errors on TAD and TCP error in AK3 and AK4 test

5. Experimental study

The proposed method was tested on a five-axis gantry machine tool as shown in Figure 5. The structure of the machine tool is described in Figure5. AK3 and AK4 tests for the comprehensive TAD and TCP error evaluation are carried out by using the DBB and R-test respectively. The test consists of two parts: one is to test AK3 twice with a double ball bar, the corresponding tool lengths are 190mm and 280mm. The other is to test AK4 twice with R-test. The corresponding tool lengths are 170mm and 260mm, respectively. The test speed is F1000mm/min.

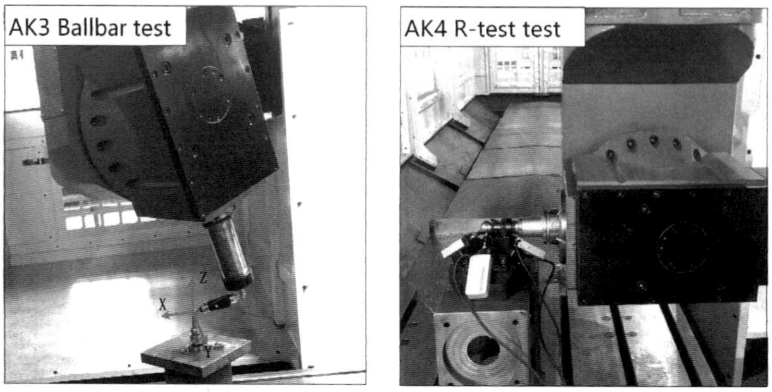

Figure 5. Experimental set up

The measurement results are shown in Table 2 and the error trajectory is shown in Figure 6. It can be seen that the maximum deviation of TAD measured by AK4 is 140 sec, and the deviation of TAD measured by AK3 is 56 sec. Combined with the analysis of rotary axis range corresponding to the two tests, the motion range of C axis of the two tests is the same, but the motion range of A axis in AK3 is [0°, 25°], and that of AK4 is [0°, 90°]. It can be seen that the difference between the two test results is mainly caused by the error of A axis.

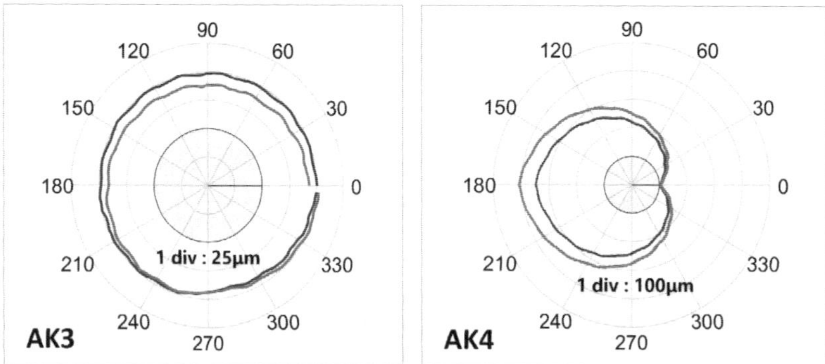

Figure 6. Contour error profiles for the AK3 and AK4 test

Table 2. TAD and TCP error measured by experiment

Error term	AK3 test	AK 4test
First TCP error	109.7 μ m	312.8 μ m
Second TCP error	89.9 μ m	251.6 μ m
TAD error	56.0 sec	140.3 sec

6. Summary

This paper presents a kinematics test and evaluation method for tool axis direction error of five-axis machine tools. Based on the AK3 and AK4 test standards in ISO10791-6-2014, a testing method and an evaluation index calculation method for tool axis direction are designed. By measuring and calculating the trajectory errors of two tool center points corresponding to different tool lengths, the TAD error can be obtained by combining the two measurements. It can effectively identify the proportion of errors caused by the deviation of tool axis in the total error of tool tip, and is conducive to faster discovery of machine tool performance problems.

7. References

[1] Tsutsumi M, Tone S, Kato N and Sato R 2013 Int. J. Mach. Tools. Manuf. 68 11-20
[2] Kenno T, Sato R, Shirase K, Natsume S and Spaan H.A.M 2020 Precis. Eng.-J. Int. Soc. Precis. Eng. Nanotechnol. 61 110-119
[3] Ezedine F, Linares J.-M, Sprauel J.-M and Chaves-Jacob J 2016 Precis. Eng.-J. Int. Soc. Precis. Eng. Nanotechnol. 43 178-186
[4] Zhong L, Bi Q and Wang Y 2017 Int. J. Mach. Tools. Manuf. 122 106-119
[5] Zhong L, Bi Q, Huang N and Wang Y 2018 Int. J. Mach. Tools. Manuf. 125 20-33
[6] Ibaraki S, Oyama C and Otsubo H 2011 Int. J. Mach. Tools. Manuf. 51 190-200
[7] ISO 10791-6:2014, Test Conditions for Machining Centres. Part 6. Accuracy of speeds and interpolations, ISO, Geneva.
[8] ISO 10791-7:2014, Test Conditions for Machining Centres. Part 7. Accuracy of finished test pieces, ISO, Geneva.

Acknowledgements
The authors gratefully acknowledge the financial support of the National Natural Science Foundation of China (Grant No. 51875357), the Advanced Technology Joint Research Fund of Shanghai Academy of Spaceflight Technology (USCAST2019-16) and the Joint Innovation Fund of China Academy of Launch Vehicle Technology (CALT201809).

IOP Conf. Series: Materials Science and Engineering **831** (2020) 012013 doi:10.1088/1757-899X/831/1/012013

Automated Driving System Disengagement Analysis and Testing Recommendations

Siyuan Liu
University of Western Ontario, Department of Electrical and Computer Engineering, London ON Canada N6A 3K7
E-mail: sissilsy@gmail.com

Luiz Fernando Capretz[1]
University of Western Ontario, Department of Electrical and Computer Engineering, London ON Canada N6A 3K7
E-mail: lcapretz@uwo.ca

Abstract. Automated Driving System refers to a vehicle system where hardware and software are collectively capable of on-road operational and tactical functions. Therefore, appropriate verification and validation procedures should be followed to mitigate unreliability and hazardousness. Autonomous Vehicle Disengagement Reports and Autonomous Vehicle Collision Reports from the Department of Motor Vehicles (DMV), California, USA are collectively used for the purpose of this research. Influencing factors are generated and consolidated from the mentioned reports and are proposed in addition to a Society of Automotive Engineers International (SAE International) standard. Stakeholders will benefit from the presented rationales and should consider the suggestive parameters throughout their developing and testing activities. This paper further recommends testing management, especially test driver management and test routes planning, for automated driving systems in accordance with the analytical results and feedback from KPMG's Global Automotive Executive Surveys.

1. Introduction

Autonomous vehicle refers to a vehicle that is capable of observing, identifying and distinguishing its surrounding static and dynamic environments, and handling, operating and maneuvering itself under unmanned conditions [1]. The technological development has shown an emerging transition from no automation to full automation along the way.

According to the Global Automotive Executive Survey for 2016 and 2017 [2] [3], the surveyed companies, including premium Original Equipment Makers (OEM) like BMW and newcomers like Google or Waymo, have increased their investments significantly in automated driving systems and self-driving vehicles compared to 2014, when the survey started to take self-driving into account. Ranked as the 9th among all 11 trends with 37% rating of importance, automated driving is not just a disruptive concept that only researchers are interested in, it has become fashionable for almost all major car makers who are putting effort into this area in order to

Dr. L.F. Capretz is currently spending his sabbatical leave as a visiting professor of Computer Science at New York University in Abu Dhabi, UAE.

maintain their status in the field as major players and also transforming their business models to service- and data-oriented models.

Given the fact that automation in commercialized and mass-produced cars serves as an essential enabling factor to urban transportation, this paper will discuss the current development of automated driving systems and recommend testing strategies. The terminology and taxonomy used in this paper are in accordance with J3016^{TM} (R) Taxonomy and Definitions for Terms Related to Driving Automation Systems for On-Road Motor Vehicles from SAE International [4]. Data about disengagements and collisions used in this paper are based those released by Department of Motor Vehicles, State of California, USA (DMV) [5] and collision reports from California Highway Patrol, State of California, USA (CHP) [6]. The recommendations presented in this paper are referring to J3018^{TM} Guidelines for Safe On-Road Testing of SAE Level 3, 4, and 5 Prototype Automated Driving Systems (ADS) [7].

2. Background

2.1. Automated Driving System

Automated Driving System, ADS, s deployed to cover a wider spectrum of functionalities and automation levels in modern vehicles. As per SAE-J3016^{TM}, ADS refers to a system of a vehicle where hardware and software are collectively capable of performing all aspects of Dynamic Driving Tasks (DDT). DDT refers to on-road operational and tactical functions which involve detection, recognition, classification and response to objects and events [4]. The National Highway Traffic Safety Administration (NHTSA) of USA is currently adopting this standard, and so as other major players in automobile industry.

2.2. Levels of Driving Automation

Based on SAE-J3016, total of 6 standardized levels should be followed by the media, manufactures, suppliers and the public. A flow chart, refers to Figure 1, is presented and it illustrates how does the automation level being determined. In the flow chart, ODD denotes Operational Design Domain which specifies those appropriate conditions for a driving automation system to work properly and OEDR is a subtask of DDT which focuses on detecting, recognizing, classifying objects and prepare to response.

2.3. Roles of Driver

As per ADS, drivers are classified into two subdivisions, expert test drivers and human drivers.

Expert test drivers are distinctively trained personnel with special skillsets and are able to safely monitor the ADS after successfully activate the system. Moreover, expert drivers need to know how to deactivate the ADS under certain circumstances especially in hazardous situations. All these actions are performed with the aid of experimental compound.

Human drivers, on the other hand, are personnel operating a particular vehicle equipped with ADS and interact with the system to perform DDT.

3. Current Development

3.1. Impact Assessment

Since the drawbacks of public transportation system do exist. For instance, people with impairment would rather stay indoor than taking a bus simply to avoid all the inconveniences. According to Canadian Survey on Disabilities (CSD) 2012, an estimated 3.8 million adults, which accounts 13.7% of total adult populations, are reported with disabilities [8]. Another factor is that the worldwide demographic shift toward aging. In 2017, about 13% of the global population which is estimated as 962 million people, are aged 60 or over. This number is projected to reach 2.1 billion by year 2050 according to United Nations (UN) [9] [10]. Despite the fact that the

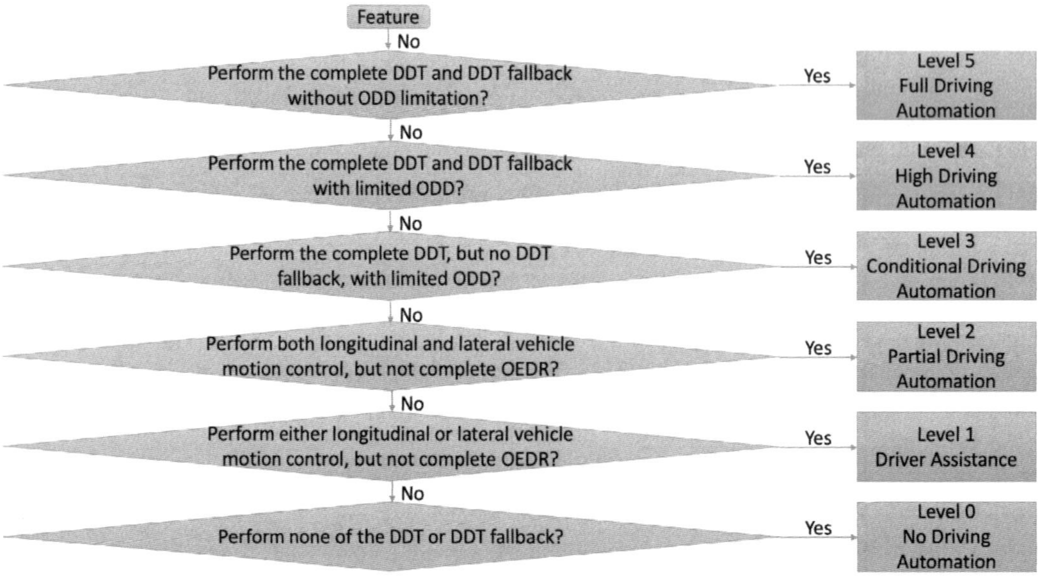

Figure 1: Flowchart for determining driving automation level

roads are also aging, the drastic increment itself is bring demands and pressures to the current transportation system. This leads to another needing factor of disrupting current automobile systems and work on something radical.

3.2. Industrial Status as of 2018

Table1 is presented with collective highlights regarding automated driving related technologies and prototypes exhibited at Consumer Electronics Show (CES) International 2018. Some of the participating companies are also permit holders with DMV and they are authorized to perform automated driving system testing on public roads in California.

It is evident that the industry has progressed significantly over the past few years, many emerging players have shown their abilities by bringing prototypes or even commercialized cars to the show, e.g. Torc. Moreover, many major premium players like BMW, Tesla and Waymo, Google's former self-driving project, have already reached a pivotal point by successfully demonstrating their capabilities in the industry.

4. Disengagement and Accident

4.1. Disengagement

As per DMV, disengagements are categorized into two types. One for automated driving with driver present, one for automated driving without driver present. For the former, from technological perspective it can be further classified as technical disengagement and non-technical disengagement, For driverless car, disengagements are accounted for the purposes of safety of the vehicle, the occupant of the vehicle and public needs.

4.2. Data from DMV, California, USA

In order to legally test prototyped motor vehicles with automated driving systems, companies need to apply for permits and report on collision and disengagement to DMV on annual basis. The reports have become available since 2015. People may think the disengagement ratio should be proportional to the miles travelled. However, there is no clear relationship between these two elements.

Table 1: Industrial Status as of 2018

Organization	Software	Year	Hardware	Year
ALMOTIVE INC	aiDrive, aiSim	2018	aiWare	2018
APTIV	Centralized Sensing Localization Planning (CSLP) platform, high-speed sensing and networking systems	2019	Work with Singapore Land Transport Authority (LTA)	2019
BAIDU USA LLC	Level 4 automation		Commercializing	2018
CLARION	Smart Cockpit Solutions			
INTEL CORP A	Intel® GO™ Automotive Development Platforms			
NAVYA INC.			Work with Keolis on autonomous robotaxis	2018
TORC	Asimov self-driving system	2007	Self-driving car	2018
TRANSDEV			Autonomous electric vehicles for public use	2018

4.3. Collision

As of April 4, 2018, a total of 63 reports were received by DMV. A collision for autonomous car epitomizes the situation when operating the vehicle on a public road causes damage to property or results in casualty. When a collision occurs, regardless of the responsibilities, the traffic authority should be informed. DMV requires companies to file Report of Traffic Collision Involving an Autonomous Vehicle, within 10 days after the accident.

4.4. Possible Causes

Based on DMV′s data, there are two types of disengagement reported, one is auto disengagement which refers to technical failure of the system and the other one is manual disengagement which means that driver has to take control of the car over the system.

Regarding the rationales, DMV has provided some indicative examples including weather conditions, road surface conditions, constructions, emergencies, accidents or collisions. Moreover, for both types of disengagements, the following root causes appear to be more frequently reported by company than those indicative causes. Unwanted manoeuvre, Perception discrepancy, Software discrepancy, Hardware discrepancy, Behaviour prediction failure and Reckless behaviours from other road users.

5. Developing and Testing Recommendations

5.1. System Development and Modification

Currently the test cars used for testing are prototypes built on existing commercial vehicles and equipped with embedded automated driving system. Commercialized vehicles refer to those cars that have already developed and are mass produced by OEMs. The electronic components in these cars are flashed with pre-defined software and such software has its own architecture.

ADS can either be flashed into existing Electronic Control Units (ECU) through CAN reprogramming or serve with an additional external hardware as add-on. Either way, interfaces between software and hardware should not collide and should not cause any conflict. The software needs to be validated and verified as per standard software testing conventions. The modules should be able to perform self-diagnostics and should be monitored all the time along with all other pre-equipped electronic modules. And prior to perform dynamic on-road testing, static stationary testing should be performed on compound and safety critical measurements should be performed.

Upon completion of necessary modifications, companies are advised to carry out an impact analysis and making sure ADS does not cause any side effects to the existing system.

5.2. Test Routes Planning

Since most of the testing activities are performed either on test tracks or public roads, many interrelated variables are needed to be taken into consideration. For instance, speed handling on wet or dry road during daytime and nighttime. If each condition is treated as one parameter, then a matrix could be formed by linking one parameter to another. Based on such matrix, almost all scenarios could be covered. However, some of the scenarios could be hypothesis and are not feasible, such scenarios could be eliminated during test routes planning.

Below are some suggestions of parameters that can be considered to build the matrix,

- *Type of road:* Freeway, expressway, driveway, intersections, roundabout
- *Timing:* Peak hour, non-peak hour, nighttime, daytime
- *Weather condition:* Dry, raining, snowing, windy
- *Season:* Spring, summer, fall, winter
- *Traffic:* Heavy vehicle, pedestrian, cyclist
- *Sign:* Speed limit, place of interest, directions, signals
- *Road condition:* Curb, slope, uphill, highway barrier
- *Location:* Car park, residential neighborhood, military base

Additionally, special testing in irregular environment should also be considered and only need to be carried out when condition permits. Testing on construction sites or testing during extreme weather conditions, these can also be added to the standard testing procedures providing that safety can be ensured.

References

[1] Autonomous car URL `https://www.techopedia.com/definition/30056/autonomous-car`
[2] KPMG 2017 KPMG's global automotive executive survey 2017 URL `https://assets.kpmg.com/content/dam/kpmg/xx/pdf/2017/01/global-automotive-executive-survey-2017.pdf`
[3] KPMG 2016 KPMG's global automotive executive survey 2016 URL `https://assets.kpmg.com/content/dam/kpmg/pdf/2016/04/kpmg-global-automotive-executive-survey-2016.pdf`
[4] SAE International 2016
[5] Department of Motor Vehicles Testing of autonomous vehicles with a driver URL `https://www.dmv.ca.gov/portal/dmv/detail/vr/autonomous/testing`
[6] The California Highway Patrol SWITRS reports URL `http://iswitrs.chp.ca.gov/Reports/jsp/CollisionReports.jsp`
[7] SAE International 2015
[8] Social and Aboriginal Statistics Division 2013 URL `http://www.statcan.gc.ca/pub/89-654-x/89-654-x2013002-eng.pdf`
[9] Ageing URL `http://www.un.org/en/sections/issues-depth/ageing/`
[10] Population Division 2017 URL `https://esa.un.org/unpd/wpp/Publications/Files/WPP2017_KeyFindings.pdf`

Research on Development and Key Technologies of Intelligent Bed-jig System for Ship Sectional Construction

Yubo Shen, Rui Wang, Mingzhi Liu, Yueting Wang, Mingshan Chen, Yaorong Zhao and Honggen Zhou*

School of Mechanical Engineering, Jiangsu University of Science and Technology, Zhenjiang, 212003, China

Syb980821@icloud.com

Abstract. Nowadays, the existing bed-jig is difficult to meet the requirements of modern ship intelligent manufacturing. This is mainly due to the following three problems: one is the redundancy of location arrangement, the other is the difficulty of moving, and the third is the discontinuity of height adjustment. For the segmental construction of ships, this paper puts forward the research and development of intelligent bed-jig system. Breakthroughs have been made in key technologies, including dynamic placement of jig and dynamic adjustment of support height. It also realizes the intelligent decision-making of process data in ship sectional construction. In general, it provides a technical basis for intelligent manufacturing of ship segments.

1. Introduction

The bed-jig is a shape membrane and workbench, which is used to make the hull surface segment and the surface three-dimensional segment. Its function is to ensure the accuracy of the ship's segment construction [1]. This needs to be accomplished in two parts: one is to support the segment to ensure the accuracy of the shape size; the other is to reduce the deformation of the segment. From the original formwork type to sleeve pillar type, flexible type and numerical control type [2], the automation level has been improved. But there are still problems. They include low level of intelligence, discontinuous adjustment of placement height, low positioning accuracy and lack of process data detection. Because of their existence, the existing jig can not meet the needs of modern shipbuilding mode. This has seriously hindered the development of the intellectualization level in the process of ship sectional construction.

In order to meet the needs of modern shipbuilding, relevant researchers at home and abroad have studied the existing bed-jig. Among them, BS Jang et al [3] proposed a method of ship sectional modeling. BWT Quinton et al. [4] introduced an analysis method of ship and marine jig under moving load. For the finite element analysis of the moving load on hull structure, the basis is provided through real-time monitoring of the positioning and load of the moving jigs. In order to provide reference for structural design and safety assessment of ship's jig, S E Lee et al [5] studied the ultimate strength of the jig through nonlinear finite element analysis. J Cao et al. [6] proposed an experimental and analytical method for a hybrid hull model. In order to realize unrestricted positioning of movable frame, V Rubino et al [7].proposed a test method of Y-type frame under distributed load and local load.

In terms of bed-jig arrangement, the research is divided into the following five parts: In order to provide theoretical basis for placement of bed-jig, Wu G.Y [8] proposed a fast algorithm to reverse the control vertex of cubic B-spline curve. In order to realize the layout design, installation and allocation of the bed-jig, Xu Peicai et al. [9] calculated the inner longitudinal wall of the bed-jig through the surface module and curved module of Tribon system. In order to allocate the placement of each segment, Xu Xing et al. [10] put forward a layout about the placement of the ship segment construction. In order to optimize the elastic support of the ship's jig support structure under impact load, Ji Yong Jin et al. [11] proposed a layout method. Wang Xiaolong [12] proposed an intelligent bed-jig, which consists of support/induction system, transmission system, control system and interactive system.

In the aspect of bed-jig body design, the research is divided into the following two parts: on the one hand, Li P.Y et al [13] invented an adjustable bed-jig and a method for automatically adjusting the support height. Firstly, the bed-jig unit is fixed on the ground or platform; Sceondly, the traction device is moved on the guide rail; finally, the height of each bed-jig is adjusted by the height adjusting device in turn. On the other hand, Chen L et al. [14] invented a movable marine bed-jig including its control system and control method. Thus the stacking yard of the marine bed-jig is reduced. And the utilization rate of shipyard site is improved. However, the control system module only puts forward the idea and lacks the relevant application.

In the aspect of construction process data acquisition, the study consists of the following four parts: Song.J.J et al [15] proposed a flexible manufacturing technology. It was based on segmented numerical control of ship's curved surface. Through post-assembly calibration and anti-deformation compensation technology, the accuracy and quality of surface forming can be guaranteed. It can effectively reduce the assembly difficulty of irregular components and improve the assembly efficiency and quality. Gu Y.F et al. [16] put forward the development of intelligent flexible system under 4.0 mode of shipbuilding industry. With the help of three-dimensional data simulation technology, automatic measurement and screw height adjustment can be realized. Kim T W et al. [17] introduced a design and manufacturing process of marine bed-jig parts. The solution optimized the assembly and welding process of the bed-jig through software data integration. Qin H R [18] and others have studied a ship intelligent painting production line. It realizes the integration of hull sectional painting data. Then it carries out dynamic analysis and prediction by real-time monitoring.

The above literature has studied the technology of bed-jig, but the existing technology cannot realize the following three functions. The first is the dynamic layout of the frame, the second is the dynamic adjustment of the support height, and the third is the effective collection and application of construction data. Therefore, aiming at the above problems, the research on development and key technologies of intelligent bed-jig system for ship sectional construction are presented.

2. Development Scheme of Intelligent Bed-jig System

According to the research and development, the main layout mode of the existing bed-jig is "welding-cutting-welding". This mode has the problems of unsustainable adjustment of support height and difficult movement. Therefore, it is not only time-consuming and laborious, but also has seriously hindered the development of modern shipbuilding mode.

In this paper, an intelligent bed-jig system oriented to ship segmental construction is developed. Thereinto, the intelligent bed-jig is the carrier of the system. And the design process of it is shown in Fig. 1. It includes three parts: one is the self-adaptive contact mechanism, which realizes the surface contact between the activating collateral head and the segmented outer plate; the other is the intelligent moving system of the bed-jig, which realizes the movement and positioning in the track; and the third is the stepless adjusting transmission mechanism, which can continuously adjust the support height of the bed-jig.

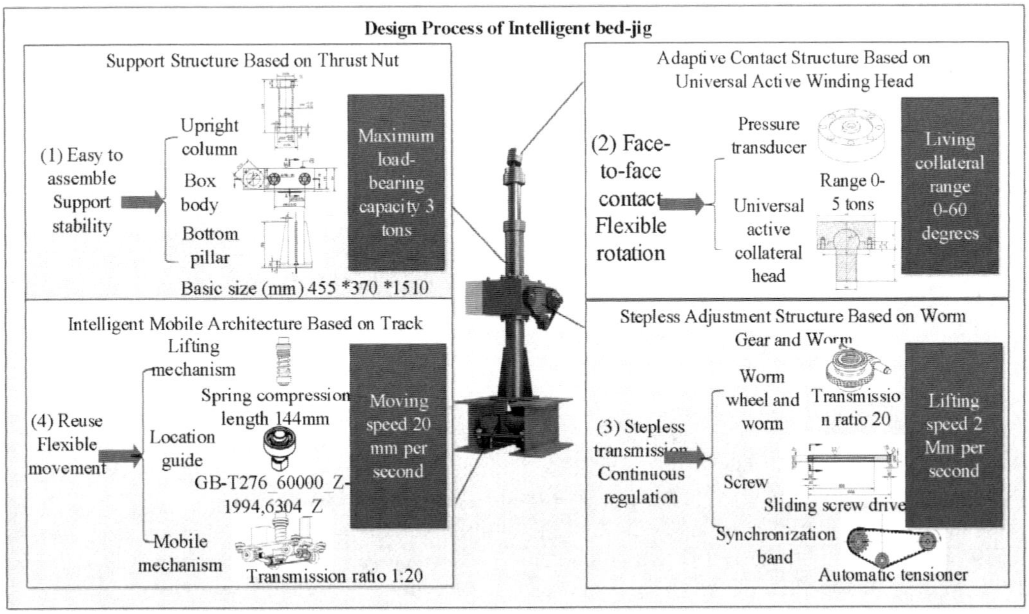

Figure 1. Design Process of Intelligent Bed-jig

As shown in Fig. 2, the adaptive contact mechanism consists of two parts: One is the support device, including a ball head and a supporting activating collaterals head, which can realize the surface contact between the jig and the ship's segmented outer panel; the other is the data acquisition module, which is mainly composed of pressure sensors for the analysis and decision-making of the control system. In order to adapt to the outer plate, the universal activating collaterals head is composed of a screw with a ball head and a supporting structure.

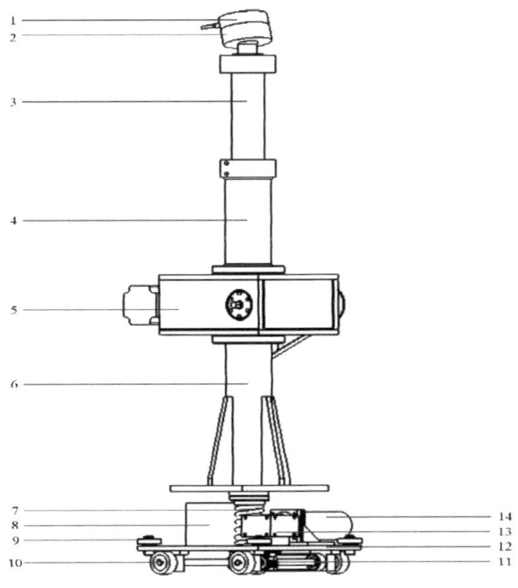

1-pressure sensor; 2-universal movable winding head; 3-lead screw; 4-upper column; 5-drive box; 6-lower column; 7-compression spring; 8-accumulator; 9-guide mechanism; 10-rubber wheel; 11-worm wheel; 12-support plate; 13-step motor reducer; 14-step motor

Figure 2. Three-dimensional model of intelligent bed-jig

Intelligent bed-jig moving system consists of moving chassis and track. The moving chassis is composed of four parts. One is the support mechanism, which is used to carry the intelligent bed-jig or lift the intelligent tire frame off the track; the other is the mobile mechanism, which is set under the support mechanism and moves along the track longitudinally; the last are the control system and positioning system, which are set above the support mechanism to precisely locate the coordinates of the placement.

Stepless regulating transmission mechanism is divided into the following two functions: Firstly, in order to realize the continuous adjustment of the support height, the stepper motor is used to output the power, and then the worm and screw are used to transfer the power; secondly, for realizing stepless adjustment of the bed-jig in the process of ascending or descending, the power is transmitted to the lead screw through synchronous belt and worm gear mechanism driven by stepping motor.

3. Key Technologies of Intelligent Bed-jig System

3.1 Dynamic placement technology of intelligent bed-jig based on multi-objective constraints

The existing bed-jig has the following two problems: the accuracy of point location and height positioning is poor, and it takes a long time to correct repeatedly. Therefore, it is difficult to improve the accuracy and efficiency of hull segmentation construction. Thus, a dynamic placement technology based on multi-objective constraints is designed, which can determine the spatial coordinates of the bed-jig quickly. Finally, the layout problem "construction-cutting-construction" was solved.

In this paper, the surface density is defined as the projection from ship segments to plane. And the distribution of mass in the projection plane is shown in Fig. 3.

The algorithm is divided into two parts: 1) Firstly, the sectional surface density model of ship is calculated by using the BOM table of the sectional components. Then the segmented centroid is solved. Finally, under the constraints of track and maximum support load, the optimal coordinates of the bed-jig are determined. 2) In the first step, discrete data points are obtained by scanning the segmented outer plate. In the second step, the surface model of the ship segmented outer plate is created by data fitting method. In the third step, the contact point coordinates of the bed-jig and the outer plate are determined under the constraints of the point coordinates and the spatial inclination angle. In the last step, coordinate conversion method is used to calculate the support height of each bed-jig.

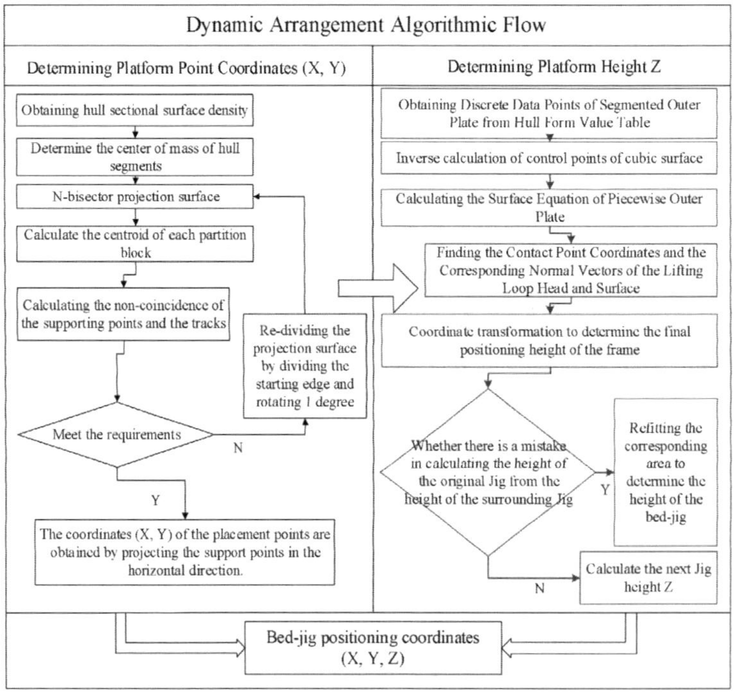

Figure 3. Dynamic Layout Algorithmic Flow

IOP Conf. Series: Materials Science and Engineering **831** (2020) 012014 doi:10.1088/1757-899X/831/1/012014

3.2 Height Dynamic Adjustment Technology Based on Measured Data

In the process of ship sectional construction, according to the shape of the segments, dynamically adjusting the bed-jig height is impossible. In order to solve this problem, this paper designs a dynamic adjusting method of the bracing height based on the measured data. This method is divided into the following three steps. Firstly, the Internet of Things (IOT) is constructed to obtain measured data. Secondly, based on the historical database and the anti-deformation rule database, the measured data are compared and analyzed. Then the deformation trend of the segmented outer plate can be predicted. And the compensation of the bracing height of the bed-jig can be calculated. Finally, the control system is used to realize the linkage adjustment of each bed-jig, so as to complete the adaptive compensation of the support height.

As shown in Fig.4, firstly, according to the measured data collected, the data management layer sets the range, including pressure value and pressure warning value. Secondly, the data management layer makes an early warning judgment on the collected pressure data. After not exceeding the early warning value, the range of comparison is made. If the collected pressure value is less than the safe range, the data management layer sends the rising signal to the PLC, and the PLC drives the stepper motor to control the rise of the intelligent bed-jig; if the pressure value is larger than the safe range, the data management layer sends the falling signal to the PLC, and the PLC drives the stepper motor to control the fall of the intelligent bed-jig. Finally, the adjusted data are stored in the database.

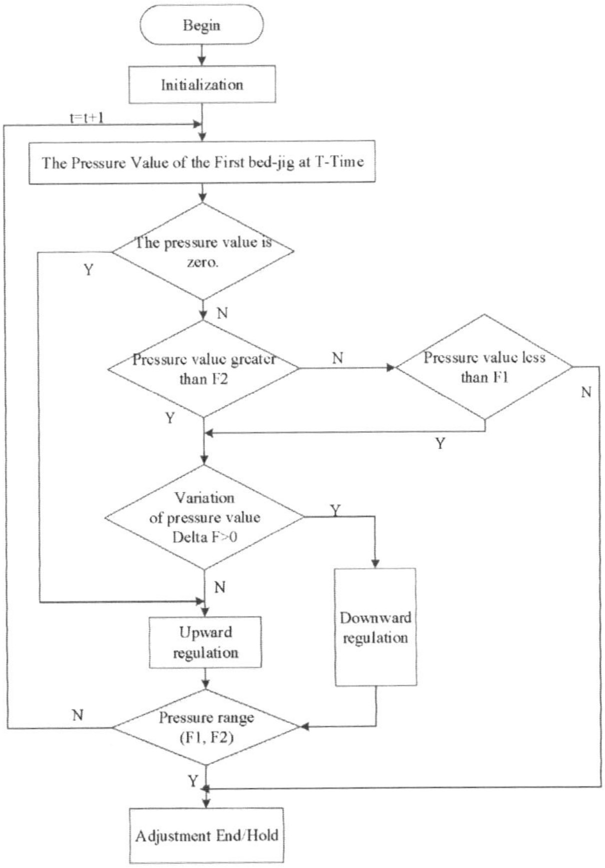

Figure 4. Height Dynamic Adjustment Process

4. Key Parts Checking

In the process of ship's segmental construction, the key parts of the bed-jig need to be checked because it will bear large external loads. There are three parts to be checked: activating collaterals head, screw

IOP Conf. Series: Materials Science and Engineering **831** (2020) 012014 doi:10.1088/1757-899X/831/1/012014

and chassis bolt. The first is that the activating collaterals head and the screw bear the weight of the ship segments. Secondly, the screw and the worm wheel are connected and driven through the thread pair, and the thread of the matching part has self-locking phenomenon. Thirdly, because the ship has a certain radian, the bed-jig support will bear the lateral force, and the reverse torque generated is borne by the chassis bolt that connects the strut to the track. Therefore, they need to be checked to ensure their stability in the process of work.

Presupposition: The weight of the ship's segmented outer plate is 10t, the intelligent bed-jig system is supported by three tracks and six bed-jigs, and the rated load of a single bed-jig is 3T. After calculation, the bearing capacity of each jig in the vertical direction is 2-3t.

4.1 Pressure Strength Checking

(1) According to the material mechanics, the checking formula of the pressure on the activating collaterals head rod is as follows:

$$[\sigma]=F/A \qquad (1)$$

In Formula 3-1, F is the pressure acting on the member, A is the cross-section area, and [σ] is the allowable stress of the material.

The material of the activating collaterals head bar is 45 steel, and the compressive strength [b] of 45 steel is 590 MPa and the yield limit [s] is 310 MPa. The calculation results show that [σ] =155 MPa.

By calculation, F≤31.2KN. The ultimate pressure of the activating collaterals rod can be calculated to be 31.2 KN. The rated working load-carrying capacity of the given bed-jig is 30000N, so the strength meets the requirements.

(2) Similarly, the cross-sectional area A of the screw is known to be 3848 mm2. Calculated by formula 3-1, F≤ 596.4KN.

The ultimate pressure on the screw is 596.4KN, and the maximum critical pressure for the whole bed-jig system to maintain stability is 846.2KN, so it meets the requirements.

(3) The bolt will be subjected to axial force Fa due to the existence of overturning moment.

$$\sigma_{ca}=4x1.3F_2/\pi d^2 \leq [\sigma] \qquad (2)$$

In the formula, F_2 is the total axial force of the screw bolt.$F_2=F_0+F_a\,[C_b/\,(C_b+C_m)]$, F_0 is residual preload. $C_b/(C_b+C_m)$ is the relative stiffness coefficient. [σ] is the allowable stress.

$F_2= 1.1Fa$, $S = 2.5$, then $[\sigma] = 128$Mpa

Then $F_2=1.1F_a \leq [\sigma]\,\pi d^2/\,(4x1.3)\ =25.06$KN, $F_a \leq 22.78$KN

From formula $F_a=M/4L_1=F_H*H/4L_1$, $F_H \leq 4F_aL_1/H=3.49$KN

The maximum lateral force that the chassis bolt can bear is 3.49KN.

Based on the above checking calculation, the following parameters are obtained:

(1) The limit pressure that the movable collar can bear is 31.2KN;

(2) The ultimate pressure that the screw can bear is 596.4KN.

(3) The maximum lateral force that the chassis bolt can bear is 3.49KN.

And they can meet the requirements.

4.2 Virtual Validation Analysis

The stress-strain analysis model of the activating collaterals head, screw and chassis bolt is obtained by ANSYS analysis.

(1) Assuming that the activating collaterals head is offset by 20 degrees and the bearing capacity is 30 000N, the maximum deformation is 1.31mm, and the maximum strain is at the junction of the activating collaterals head rod and the sensor.

(2) Loading 30000N pressure on the top of the screw, the maximum deformation is 0.041mm, and the maximum stress is 37.8KN, which meets the application requirements.

(3) Assuming that the contact between the bolt and the bottom surface is bonded, the nut and the bottom surface of the track are bonded, and the rough between the bottom and the track. The theoretical calculation shows that the pretension limit is 18.224KN, the maximum transverse force on the chassis is 3490N, and the maximum deformation is 0.103mm. The stress at the beginning of the thread is the largest, which conforms to the actual situation.

In summary, the activating collaterals head, screw and chassis bolt can meet the design requirements. And the stability of the bed-jig system is guaranteed in the process of ship's sectional construction.

5. Conclusion and future works

In this paper, an intelligent bed-jig system for ship sectional construction is developed. And a kind of movable intelligent bed-jig is studied, which has self-adaptive contact mechanism and stepless regulating transmission mechanism. The next step is to optimize the flow chart of dynamic placement algorithm and the structure of intelligent mobile system. At the same time, the function of intelligent bed-jig system will be fully utilized and deeply developed, so that it can also play a role in the process of ship sectional assembly and welding.

6. References

[1] Xu Z K. Shipbuilding Technology [M] Beijing: China Communications Press. 2010.
[2] She J G, Gu Y F. Analysis and Design of Flexible Bed-jig Based on Center Manufacturing Ship [J]. Ship Science and Technology20, 08, 30(1):120-123.
[3] BS Jang, YS Suh, EK Kim, et al. Automatic FE modeler using stiffener-based mesh generation algorithm for ship structural analysis[J]. Marine Structures,2008,21(2-3):294-325.
[4] BWT Quinton, CG Daley, RE Gagnon, et al. Guidelines for the nonlinear finite element analysis of hull response to moving loads on ships and offshore structures[J]. Ships and Offshore Structures, 2017, 12 (supl): S109-S114.
[5] SE Lee, AK Thayamballi, JK Paik.Ultimate strength of steel brackets in ship structures [J].Ocean Engineering, 2015, 101:182-200.
[6] J Cao, JL Grenestedt, WJ Maroun. Testing and analysis of a 6-m steel truss/composite skin hybrid ship hull model[J]. Marine Structures,2006,19(1):23-32.
[7] V Rubino, V S Deshpande, NA Fleck. The collapse response of sandwich beams with a Y-bed-jig core subjected to distributed and local loading [J]. International Journal of Mechanical Sciences, 2008, 50(2):233-246.
[8] Wu G Y, Wang X H. A Fast Algorithms for Inversely Finding the Control Vertex of Cubic B-Spline Curve [J]. Journal of Hangzhou Dianzi University, 2005, 25(3).
[9] Xu P C, Hong A C. Application of TRIBON System in Calculating Inner Longitudinal Wall Platform [J]. Guangzhou Shipping Science and Technology, 2012, 32 (02):54-58.
[10] Xu X, Gu X B. Bit Layout Plan and System Development for Segmental Construction of Ships [J]. Shipbuilding Technology, 2014(2).
[11] Ji Yong Jin, Kwak Jeong Seok, Lee Hyun Yup, et al. Optimal Arrangement of Resilient Mount installed on Bed-jig Support Structure at Shipboard Equipment under Shock Load[J]. Journal of the Society of Naval Architects of Korea, 2015, 52.
[12] Wang X L. Technology and Development Prospect of Ship Bed-jig [J]. China Water Transport (Second Half of the Month), 2015, 15(03):18-20.
[13] Li P Y. A Method of Automatic Adjustment of Support Height of Shipbuilding Bed-jig and Adjustable Bed-jig: China. CN201810115290.2 [P] 2018-02-06.
[14] Chen L, Meng Q etc. Control System and Control Method of Movable Marine Jig and Marine Jig: China. CN201810712928.0 [P] 2018-07-02.
[15] Song J J,Yu Y,Wang J,Chen Y,Zhao S J. Research on Flexible Manufacturing Technology of Numerically Controlled Jig Based on Surface Section of Ship[J].Ship and Marine Engineering, 2016,32（02）: 63-68.
[16] Gu Y F, Xie R. Development of Intelligent Flexible Bed-jig System in Shipbuilding Industry 4.0 Mode [J]. Jiangsu Ship, 2016, 33(3):29-31.
[17] TW Kim,SS Lim ,HH Seok,et al.Concurrent engineering solution for the design of ship and offshore bracket parts and fabrication proess[J].International Journal of Naval Architecture and Ocean Engineering,2013,5(3):376-391.
[18] Qin H R,Wang Y,Zhong Q W,Dong H. Research on Intelligent Coating Production Line for Ships [J]. Marine Engineering Equipment and Technology, 2018, 5(04):288-292.

IOP Conf. Series: Materials Science and Engineering **831** (2020) 012015 doi:10.1088/1757-899X/831/1/012015

Flow Analysis of Two-Phase Flow in Curved Pipe of Polar Cruise Seawater Pipeline

Zongqiang Liu, Li Xu*, Guanchen Liu and Cheng Tao

School of Energy and Power Engineering, Wuhan University of Technology, Wuhan, China

xuli@whut.edu.cn

Abstract. In order to study the flow characteristics of seawater-ice crystals two-phase flow in the polar cruise seawater pipeline system, the curved pipe in the pipeline system is taken as the research object. Through numerical simulation, the effects of different initial conditions on the flow of seawater-ice crystal two-phase flow in curved pipe are analyzed, and the flow analysis results are obtained. The results show that in the curved pipe, the fluid is separated at the bend, and the pressure loss at the corner of the curved pipe is the most serious; the increase of the flow rate and the IPF (ice packing factor) will cause the ice crystals to gather toward the center; when the fluid has a high IPF, the ice crystals can be converted into a symmetric distribution state by being concentrated on the inner wall surface of the pipe at a relatively low flow rate; in the range of 0.5 m/s-2 m/s, the increase in fluid velocity has a significant effect on the distribution of ice crystals, and when the fluid velocity exceeds 2 m/s, the increase in flow rate has no significant effect on the distribution of ice crystals.

1. Introduction

As the trend of economic globalization continues to deepen, people's demands for life and values have changed greatly. Eco-tourism has emerged as the times require. The most fascinating snow and ice scenery are the most fascinating for tourists [1]. As the demand for the global adventure tourism market continues to increase, small-scale expedition cruises will enter the market in the next few years. To date, at least 11 cruise lines around the world have announced plans to build new cruise ships. By the end of 2021, a total of 22 new expedition cruise ships will enter the market [2].

When the polar cruise is sailing in the polar region, due to the extremely low ambient temperature, there are ice floes in many navigation areas [3]. Some small ice crystals may flow into the seawater pipeline system. The presence of ice crystals will inevitably affect the seawater pipeline system [4]. Will lead to the failure of the ship's power system [5]. Therefore, it is necessary to carry out the analysis of the flow of seawater-ice crystals two-phase flow in the polar cruise seawater pipeline system.

2. Mathematical model

In the seawater-ice crystals two-phase flow, the Euler-Euler two-phase flow model based on particle dynamics theory is used. The Euler-Euler model uses the liquid phase and the solid phase as the continuous phases of interaction to establish their corresponding N-S equations. In numerical simulation, the flow of two-phase flow is regarded as turbulent flow and incompressible flow. The governing equations have continuity equation, momentum conservation equation, energy conservation equation. The RNG turbulence model takes the solid-liquid mixed phase as the research object, and comprehensively considers the influence of the liquid phase and the particle phase on the flow process.

3. Numerical procedures

3.1. Thermal properties of seawater and ice crystals

In the range of 0-20m in depth, the salinity of seawater is about 11‰. For the convenience of calculation, the salinity of seawater is 10‰. The temperature of the seawater-ice crystals two-phase flow in the seawater system pipe is about 0 °C. When the salinity is 10 ‰ and the temperature is around 0 °C, the density and specific heat of seawater do not change significantly with temperature, which can be regarded as a fixed value. The thermal conductivity and viscosity of seawater vary significantly with temperature and cannot be considered as a fixed value [6].

Table 1. Seawater and ice crystals physical properties at 0 °C

Phase name	Density $(kg \cdot m^{-3})$	Specific heat $(J \cdot kg^{-1} \cdot K^{-1})$	Thermal conductivity $(W \cdot m^{-1} \cdot K^{-1})$	Viscosity (Pa·s)
Seawater	1003	4140	0.56	1.82×10^{-3}
Ice crystals	900	2080	2.21	-

3.2. Model building and meshing

A spatial 3D model of the 90°curved pipe was built in Solidworks software, and then imported into the ICEM software for meshing. Due to the simple structure of the curved pipe, the structure grid which is suitable for simple geometric structure and high quality is selected, determine the number of meshes is 89100and the quality is above 0.65. Figure 1 shows the meshing of the 90°curved pipe.

Figure 1. Grid Division Map

3.3. Flow analysis in curved pipe

The fluid velocity is set to 0.5m/s-3m/s, the IPF is set to 5%-30%, the ice crystals diameter are set to 5×10^{-4} m and the diameter of the curved pipes are 0.04 m. The two-phase flow model uses the Euler-Euler model to set the boundary condition and solution methods in the software. Considering gravity, the direction of gravity coincides with the negative direction of the Y-axis. The pipe flow simulation does not consider the influence of phase change on the flow. The convergence accuracy is set to 10^{-4}, and the wall surface adopts the fixed wall temperature condition.

In Figure 2, it can be clearly observed from the figure that the pressure loss at the corner is the most serious, and the main reason for this phenomenon is the energy loss caused by the secondary flow phenomenon. The secondary flow means that the lateral pressure exerts a certain effect on the flow along a certain boundary, so that it has a force parallel to the boundary, so the fluid velocity near the boundary is lower, and the fluid is more offset than the fluid farther from the boundary, this caused a secondary flow.

In Figure 3, when the seawater and ice crystals mixture flows turbulently in the curved pipe, the fluid movement is hindered by the deformation at the bend, and the velocity of the flow layer near the wall

IOP Conf. Series: Materials Science and Engineering **831** (2020) 012015 doi:10.1088/1757-899X/831/1/012015

is lower at the outer side of the pipe, resulting in less centrifugal force. However, the flow rate in the main flow area is large, and the centrifugal force exerted by the fluid is also larger, so that the main flow area flows toward the near wall surface, and additional flow is added to the fluid. At the same time, in order to maintain the mass conservation in the pipeline, the fluid also forms a flow of the wall area to the main flow zone.

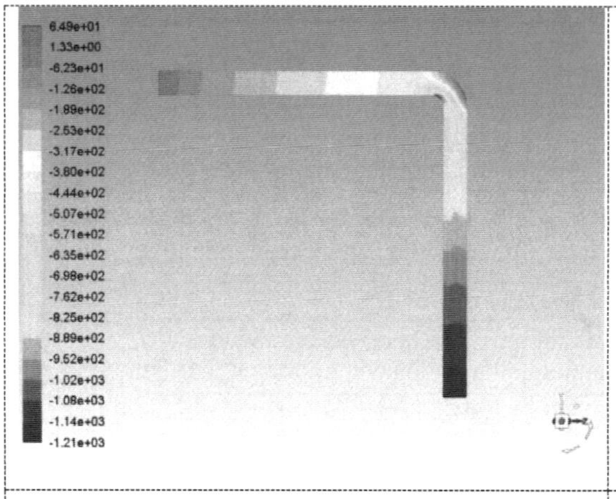

Figure 2. Curved longitudinal section pressure distribution cloud map

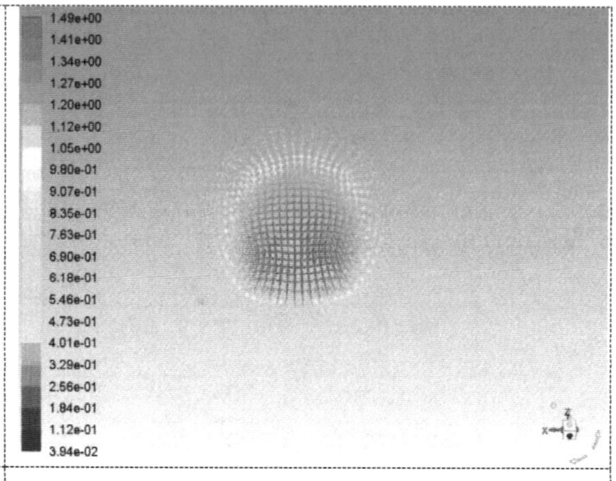

Figure 3. Velocity change of ice crystals in longitudinal section

In Figure 4, the flow rate of the ice crystals continues to increase as the fluid does not reach the bend. The ice crystals near the inner wall accelerate significantly when they are closer to the bend. When passing through the bend, the flow rate of the ice crystals exceeds the speed of the nearby seawater. Due to the centrifugal force, the ice crystals flow toward the wall surface of the pipeline, and finally pile up on the outside of the pipeline.

In Figure 5, when the flow rate of the ice crystals is sufficiently large, the centrifugal force is also large, the fluid will have a tendency to move toward the wall area, and the boundary layer is separated at about 45° of the curved section. As the ice crystals continue to flow downstream along the curved section, the separated boundary layer will reattach to the tube wall, thus reattaching points appear on the tube wall. In the vicinity of the upstream and downstream of the reattachment point, ice crystals near the inner wall flow in opposite directions.

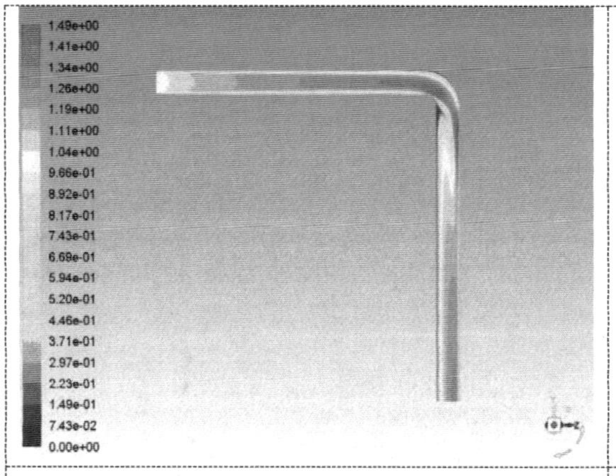

Figure 4. Ice crystals grain velocity distribution cloud diagram of longitudinal section of curved pipe

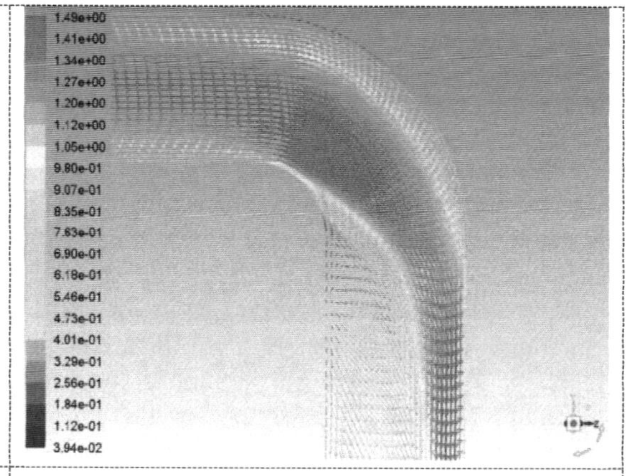

Figure 5. Curved pipe longitudinal section ice crystals velocity vector cloud illustration

IOP Conf. Series: Materials Science and Engineering **831** (2020) 012015 doi:10.1088/1757-899X/831/1/012015

3.4. Distribution of ice crystals in curved pipe

3.4.1. Effect of different import speeds on the distribution of ice crystals. The numerical simulation shows the distribution of ice crystals and seawater in the cross section of the straight pipe section of the curved pipe which is 200mm away from the inlet. The simulation results are imported into the TECPLOT software for processing, as shown in Figure 6. (a) to (f) show that in the case of IPF=10%, the ice crystals volume distribution cloud map inside the pipeline when the fluid velocity are 0.5m/s,1m/s,1.5m/s,2m/s,2.5m/s,3m/s.

It can be concluded from Figure 6 that when the fluid flow velocity is small, since the density of the seawater is greater than the density of the ice crystals, the ice crystals will flow toward the wall area of the pipe under the action of the floating force. As the flow rate increases, the turbulent flow energy of the fluid also increases. Under the action of turbulence, the ice crystals will obtain a large buoyancy, which is weakened by the influence of gravity, and the ice crystals are evenly distributed in the fluid.

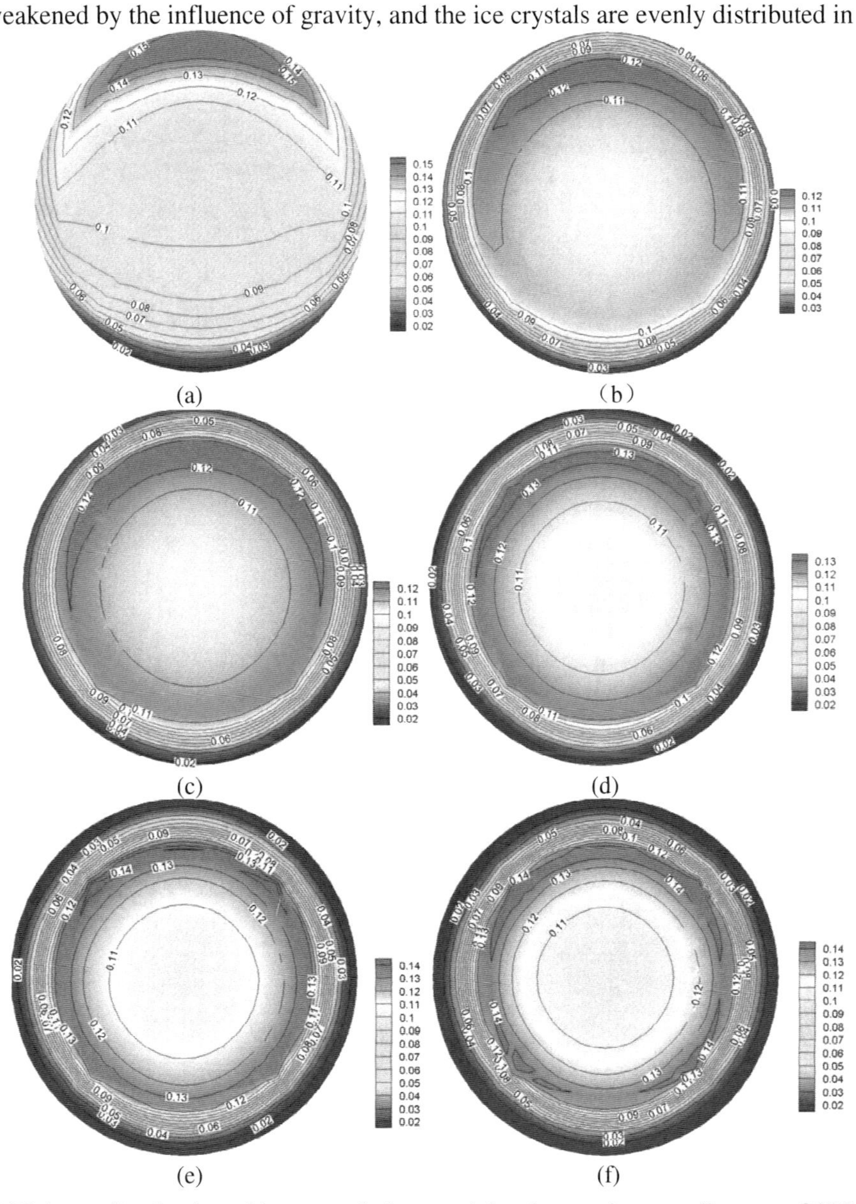

Figure 6. Volume distribution of ice crystals in a straight pipe section at a distance of 200 mm from the inlet of the curved at different seawater flow rates.

When the flow rate is 2m/s, the fluid in the pipeline exhibits a symmetric suspension flow regime, and the ice crystals near the wall region are less than the mainstream zone; When the fluid velocity exceeds 2 m/s, the outer volume of the central portion of the pipe is more uneven than the inner ice crystals volume. In the outer area of the center of the pipeline, fluid turbulence does not affect the particles, the suspension of ice crystals is mainly affected by the collision between the particles and the lift of the particles, the ice crystals appear unevenly distributed; The volume of ice crystals near the wall of the pipe. The fraction is relatively small, because the ice crystals will collide with the inner wall of the pipe near the inner wall, and then the ice crystals will flow toward the central zone.
With the increasing flow rate, the degree of collision of ice crystals with each other will increase and increase, and its effect on the suspension of ice crystals is also increasing. At this time, the volume fraction distribution of ice crystals does not change significantly.

3.4.2. Effect of different ice content on ice crystals distribution. In Figure 7, (a) IPF=30%, fluid velocity V=1 m/s; (b) IPF=10%, V=1 m/s. As can be seen from the figure, when the ice content of the fluid increases, the ice crystals in the pipe tend to be symmetrically distributed. This is because the increase of the ice-containing rate exacerbates the collision between the ice crystals, which generates a suspension force for the ice crystals to provide lift, and weakens the ice crystals by gravity. (a) is more symmetrical than (b), because the collision of the particles with each other is more severe in the case where the volume fraction of the ice crystals is higher. (c) For IPF=10%, V=3m/s, it can be seen that at higher flow speeds, more ice crystals flow to the outer wall surface of the curved section, because the centrifugal force is stronger, twice the effect of the flow at higher flow rates is more pronounced.

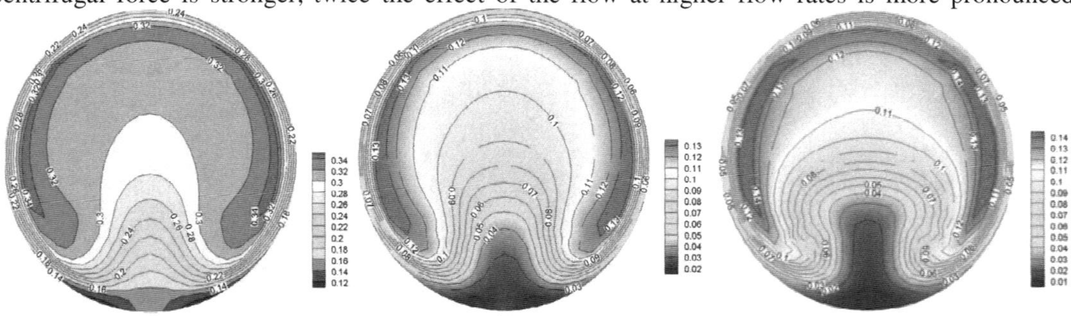

Figure 7. Volume distribution of ice crystals in different ice rates and flow rates at the curved.

4. Conclusions

The flow characteristics of two-phase flow in the curved pipe of the Polar Cruise Seawater Pipeline System are discussed. Concluded as follows:

(1) In the curved pipe, the fluid is separated at the curved, and the fluid in the near wall area will flow toward the inner wall surface of the pipe, causing a secondary flow phenomenon.
(2) The pressure loss at the corner of the curved pipe is the most serious.
(3) The increase in flow rate and ice rate will cause ice crystals to aggregate toward the central.
(4) In the 0.5m/s-2m/s, the increase of fluid velocity has obvious influence on ice crystals distribution. When the fluid velocity exceeds 2m/s, the increase of flow velocity has no obvious effect on ice crystals distribution.
(5) when the IPF increases, the ice crystals in the pipe tend to be symmetrically distributed.

5. References

[1] Wang Zipan,*Discussion on the Development of Antarctic Tourism in China.* Tourism Journal, issue. 06, pp.95-100, 2003.
[2] Fan Jun, *Adventure Cruise, the next market hotspot.* China Port, issue. 04, pp. 33-38, 2018.
[3] Wang Qingyuan,Li Qingquan,Wang Lanning,*Numerical Simulation of the Influence of Side Boundary Melting on Arctic Sea Ice.* Chinese Journal of Polar Research, vol.25, issue. 1, pp.84-89, 2013.
[4] Tian Chuan,*Overall design and implementation of polar marine environmental monitoring system.* Ocean University of China, Qingdao.2009.

[5] Wang Zhongchen,*Ship Power Plant.* Dalian Maritime University Press. Dalian, 2013.
[6] Melinder A, Ignatowicz M,*Properties of seawater with ice slurry use in focus.* International Journal of Refrigeration, vol.52, pp.51-58,2015.

Acknowledgment
This work is financially supported by the National Natural Science Foundation of China (Grant no.51479152), projects from Key Lab. of Marine Power Engineering and Tech. authorized by MOT (KLMPET2018-01) and the Fundamental Research Funds for the General Universities (WUT: 2019III 046GX).

Conceptualising an Inspection Robot for Tunnel Drainage Pipes

G Ecker[1], B G Zagar[1], C Schwab[2], F Saliger[3], T Schachinger[3] and M Stur[4]

[1] Institute for Measurement Technology, Johannes Kepler University Linz, 4040 Linz, Austria
[2] ASFINAG Service GmbH Süd, 8074 Graz, Austria
[3] ÖBB-Infrastruktur AG, 1020 Vienna, Austria
[4] Sachverständigen Büro für Boden + Wasser GmbH, 4210 Gallneukirchen, Austria

gabriel.ecker@jku.at

Abstract. For inspection purposes first ideas of an in-pipe robot are presented. The idea of a rover with respect to the environmental conditions, as well as details of structure, actuation, measurement and control concepts is discussed in this document. It should be able to navigate through rather narrow and buckled drainage pipes with diameters not much larger than the robot itself, so the design is aiming for a most compact and potentially self-disassembling solution necessary for cases of stuck robots.

1. Introduction

The problem of scaling, consisting of mostly calcite [1], [2] in drainage systems, of both road and railway tunnels, has been noticed in water pressure-relieved tunnels all around the world [1], [3], [4] reducing the in-pipe water flow and clogging the drainages. Clogged drainage pipes can cause breakdown of tunnel operation due to water ingress [4] or can even endanger the tunnel structures [4], [5]. To avoid such situations, the tunnel drainage system needs to be maintained regularly. At the moment, drainage cleaning represents the largest part of all regular maintenance work that is necessary and is causing high costs [5], [6].

Under the current predetermined maintenance regime, tunnel drainages are cleaned at fixed intervals mostly, which either depends on experience or is determined by opportunities of track availability, budget, etc. [6]. So the idea of determining the location and thickness, as well as the removal resistance of calcite and other sediments, independent of the tunnel traffic and using this information to improve the cleaning process is obvious. Beside flushing nozzle-based cameras, there are preliminary studies of measurements using stationary sensor systems of various kinds [6], integrated into newly built drainage pipes, with moveable inspection devises or by estimating the in-pipe conditions out of the surveyed data, like drainage water or scaling properties as well as geologic facts and experience [7].

Such information is currently raised in a uniform inventory of the conditions of Austria's railway tunnel drainage system, as part of the Task Force Drainage (TFD), allowing estimations on the cleaning intervals and location [7]. Another objective of the TFD is to minimize the in most cases necessary closures of the affected tunnel when inspection and maintenance work is done [7]. A kind of robot, including sensor systems, able to drive through the drainage pipe system, may allow a

measurement based inspection, providing additional information of the in-pipe conditions, without effecting the tunnel operation.
There are already existing robots for oil and gas pipelines [8], but their physical dimensions differ significantly from those required in drainage systems. Conventionally, these are steered manually, enabled through cameras [9]. The evaluation of the recorded video material is then done by inspection. Therefore, an autonomous robot is desirable, particularly considering the total length of tunnel drainage pipes that might easily be in the order of hundreds of kilometres [6].
In summary, the objective of this project is to develop a robot for inspection purposes only, capable of navigating independently through narrow drainage pipes, passing kinks, junctions and half-open tubes. Thereby also recording the position and measuring the thickness of sediments spatially resolved.

2. Locomotion strategy

2.1. Environmental conditions
The study is for typical drainage pipes with 250 mm outer diameter with a maximum reduction of 20 mm by scale deposits. Kinks up to 30° and man holes with half-open tubes must be navigable. Additionally, junctions to maintenance pipes for cleaning are encountered regularly and must be resolvable. The robot should be robust against the conditions in the pipes (dust, air humidity, alkaline water, various water levels and currents) and should need little maintenance [6].
The technical properties of the pipes may not be negatively changed by the, even many times, passing rover and the inspection must not impair the operation of the tunnel, anyway [6].
As a result of the long distances, in the order of the tunnel lengths, to be covered in the drainage system the robot cannot be powered tethered via cables.
Locomotion should be possible in both directions, so that in case of blocked pipes the rover can move out in reverse. If the robot gets stuck or totally fails, pressurised water must be able to destruct it into manageable pieces to be washed out easily. Cleaning operations by the robot were explicitly not defined as a goal in the project, since there is currently no possibility of removing the loose scale deposits after flushing by an autonomously operating robot.

2.2. In-pipe locomotion
As consequence of possibly half-open pipe sections, pipeline inspection gauge (PIG) (Figure 1A), screw (Figure 1D) and inchworm robots (Figure 1F) are unsuitable. Furthermore propeller driven rovers (Figure 1G) cannot be used, since no minimum water-level can be guaranteed. Walking concepts (Figure 1H) are generally complex, quite large and slower than other types [10], leaving wheeled, tracked and snake robots as suitable locomotion strategies.
Because of the advantage when generating the necessary friction of tracked locomotion [10], [11] and reduced effort to control compared to snake type [11], a robot concept with four tracks allowing wall-pressing for increased friction [11] is chosen. Furthermore, a rover with quadruple support is able to also pass the aforementioned half-open pipe sections without the risk of toppling over.

2.3. Adjusting mechanism
Passing kinks, junctions and half-open tubes is definitely challenging. To overcome these problems, a kind of deformable robot is advantageous. Based on the mechanics from [12], an adjusting mechanism, described in section 5.1., is designed. The combination of linear actuation and spring loaded tracks allows adapting the robot to the pipe's inner diameter and enables to generate additional friction if deemed necessary. This degree of freedom may permit driving from half-open tubes into closed ones by reducing the rover's size and adapt it again in the closed pipe section.

2.4. Passing elbows
Tracked type robots with wall-pressing technique require a differential drive to allow them to corner, this can be done with individual steerable motors or gear mechanisms [8].

IOP Conf. Series: Materials Science and Engineering **831** (2020) 012016 doi:10.1088/1757-899X/831/1/012016

When the rover needs to pass an elbow its outer diameter has to decrease, springs and adjusting mechanism are allowing such changes. Figure 2 shows the worst-case scenario in an 30° elbow. With the geometric relations

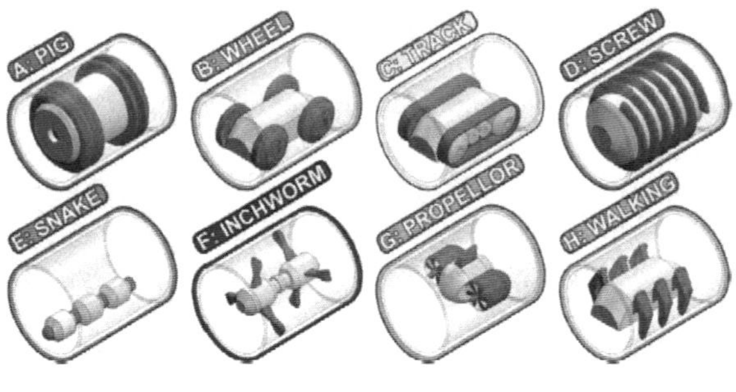

Figure 1. The eight main elements of in-pipe locomotion [10].

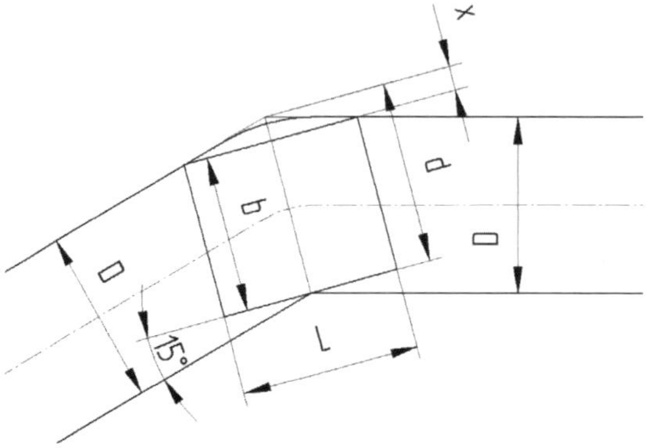

Figure 2. Situation in an 30° elbow

$$d = \frac{D}{\cos(15°)}, \tag{1}$$

$$x = \frac{L}{2}\tan(15°), \tag{2}$$

$$b = d - x = \frac{D}{\cos(15°)} - \frac{L}{2}\tan(15°), \tag{3}$$

the minimal size of the robot can be calculated from the track's length and pipe's inner diameter. This leads to a maximum outside dimension b of about 206 mm, calculated with a pipe diameter D of 230 mm and a track length L of 240 mm.

3. Basic features

Figure 3 gives an overview of the basic construction of the robot. Its main parts are four track units (components A), connected through four spring assemblies for each track unit (components C) with the skeletal structure, including linear bearing and stepper motor (component B).
The overall length of the robot is 414 mm, the track length is 240 mm and its square outline can vary between 173 mm and 330 mm. It is able to pass 30° elbows as described in section 2.4. According to the calculation of the used CAD software the rover weighs slightly over 5 kg.

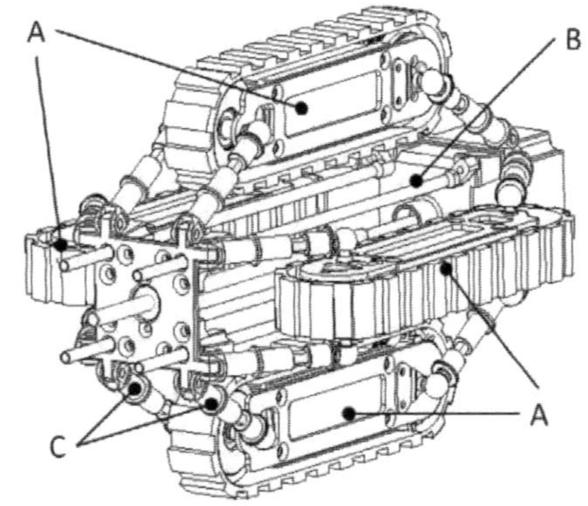

Figure 3. 3D CAD model of the robot.

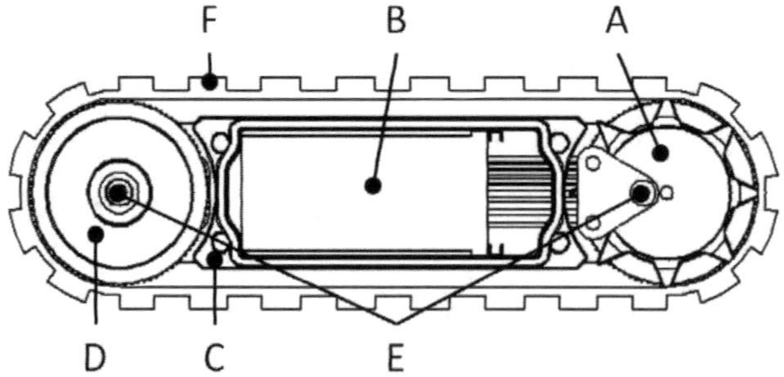

Figure 4. Components of a track unit (case opened).

4. Track unit

Figure 4 shows the components of one track unit. A: Brushless DC motor with mounted sprocket-wheel, B: Battery pack, C: Waterproof case, D: Passive sprocket-wheel, E: Ball bearings, F: Track. They are all mounted via two stainless steel supports, one on each side.

4.1. Actuation

An efficient actuation system, including rotary encoders as well as the energy storage must be devised. Narrow pipe diameters require space-saving designs wherever possible, therefore and also because of the humid environment a classic brushed motor with a then necessary gearbox is not the most appropriate choice. With a brushless DC outrunner motor direct actuation and easier corrosion protection are possible.

As a consequence, as main actuator the brushless DC motor *DF45L024048-A2* [13] is used. It is an outrunner motor with a rated power of 65 W that includes three hall sensors [13] to facilitate its control. Tests showed that these hall sensors are absolutely necessary when driving such motors at low speeds and noticeable torque without field oriented control.

Due to space constrains the sprocket-wheel is glued onto the rotor and its star like extension meshes into the track, preventing the motor from slipping through. The motor contains iron as ferromagnetic material that easily corrodes when subjected to water. Therefore, all parts of the motor are covered with epoxy resin and the bearings are replaced with non-corroding ones, this allows a wet operation, i.e. water is allowed to penetrate into the motor.

Another advantage of a brushless DC motor compared to a brushed one is the possibility of freewheeling in cases of failure or when the motor is switched off. In return, a more complex motor controller is required and the cost is higher.

The battery pack contains four 18350 lithium ion cells, resulting in a 14.8 V, 1100 mAh accumulator. A battery management system (BMS) is limiting the maximum current drain to 7 A.

To save energy while in motion one or two tracks can be turned into passive mode, depending on how much feed force is needed and in which orientation or pipe section the rover is operating in. Anyway, the battery's state of charge must be monitored to avoid unplanned standstills in the tubes.

To determine the robot's position, the distance covered is measured by counting turns of the tracks, when at least one of them is set to passive mode the accuracy is even higher, since the belt cannot slide through accidentally.

4.2. Electronics

As a consequence of the limited space in the pipe a compact controller is required. Furthermore, movable cable connections are to be avoided in order to not hamper waterproofness. Therefore, it is obvious to choose surface mounted devices (SMD) for the brushless controller. To avoid movable cables and to avoid plugging when track units are changed Wi-Fi connected microcontrollers seems to be a good choice.

Verification of the prototype track unit (see section 4.4.) showed that a *DRV8306* brushless controller in combination with three *CSD88584Q56DC* half bridge drivers, all from *Texas Instruments©* provides sufficient drive power even at low rotation speeds.

An *ESP8266* microcontroller with integrated Wi-Fi is used to implement an overlain regulator circuit. The microcontroller also ensures the communication with the main controller and feeds back measured data, like the number of turns, the state of charge of the batteries and errors.

The electronic circuit is designed to limit the motor current to 6 A resulting in a maximum peak motor power of approx. 90 W. So overall the rated drive power is 260 W (4 x 65 W), with peaks up to 360 W (4 x 90 W). The estimated mean drive power depends strongly on the applied normal force, determining the friction force, as well as the water current situation and is expected to not exceed 30 W in the mean, resulting in a minimum operating time of approx. 2 hours.

4.3. Mechanics and materials

The humid and alkaline environment conditions, typically encountered in drainages and the demand for weight minimization require certain materials to be chosen. In this context plastics for low mechanical loads and stainless steel for higher loaded components are suitable. Therefore, the cases as well as the sprocket-wheels are made from plastic and the bearings, ball pins, screws and nuts are made from stainless steel.

IOP Conf. Series: Materials Science and Engineering **831** (2020) 012016 doi:10.1088/1757-899X/831/1/012016

The spring assemblies connected through ball joints press the track units against the drainage wall which in combination with the silicone rubber material of the tracks gives sufficient friction even on wet pipe walls.
To get the whole case containing all electronics and the battery pack waterproof an O-ring is used and the feed through for cables is sealed with epoxy resin.

4.4. Prototype
Figure 5 shows a photo of one track unit, still with external electronics, used for the verification of the actuation system. It also contains the battery pack powering the whole system. The spring assemblies are also mounted, but not yet attached to the skeletal structure. This prototype is able to read in the desired rotational speed via Wi-Fi and controls the motor speed with a PI-regulator. While testing the unit it turned out that the reachable torque at desired drive speeds as well as the forward forces when pressing the track against a wall are more than sufficient for the assigned task, while the motor current stays well below the maximum of 6 A. Therefore, this particular controller is used further on.

5. Skeletal structure
Figure 6 shows the components of the skeletal structure. A: Front case, containing stepper motor and electronics, B: Back case, containing battery pack and the rotary encoder, C: Linear bearing, D: Lead screw, E: Stops, sliding bearing, F: Connecting sheets.

5.1. Mechanics and motion
To drive the adjusting mechanics a corrosion-resistant and power-saving linear actuation, which is able to withstand considerable axial forces, allows self-disassembling in case of failure and holds the track units in position, is needed.
A lead screw made from stainless steel, actuated by a stepper motor seems suitable. Due to the low angle α (Figure 6) there are considerable axial forces to be expected. Therefore, an additional bearing, accepting axial loads, needs to be installed. To prevent it from blocking there are slide bearings placed inside the stops. Moreover the stops, including the back case, avoid contact of two tracks.
Both cases as well as the stops are made from plastic to save weight and to allow for Wi-Fi connections. The waterproofing is again done with O-rings. For the lead screw shaft seals are used.

5.2. Electronics
As counterpart to the Wi-Fi connection, the data storage, the control of the stepper motor and for numerical processing as well as measuring purposes another microcontroller is used. This main controller again is an *ESP8266* microprocessor, controlling the stepper motor via a *MP6500* stepper driver from *Pololu*©. A rotary encoder allows determining the distance between the connecting sheets. On the one hand, this is necessary for the calculations, on the other hand, this allows getting feedback of the translational position, so that the motor current can be adapted and end positions can be set.
All electronics in the skeletal structure are powered by a two cell lithium ion battery pack located in the back case (Figure 6 component B). The BMS limits the current to 2 A. The lead screw was designed to be self-locking, so no holding torque is required.

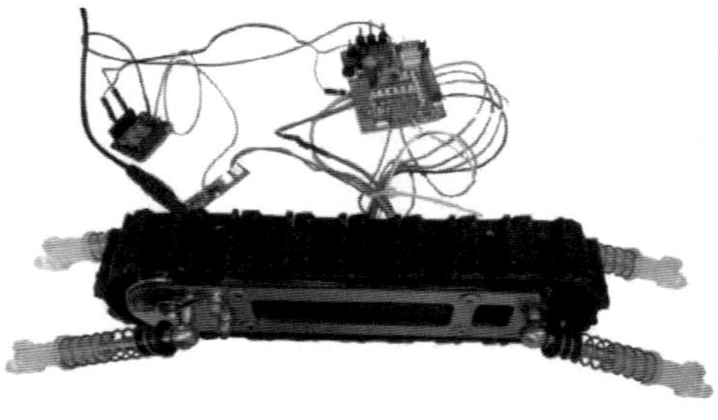

Figure 5. Track unit prototype with electronics separated.

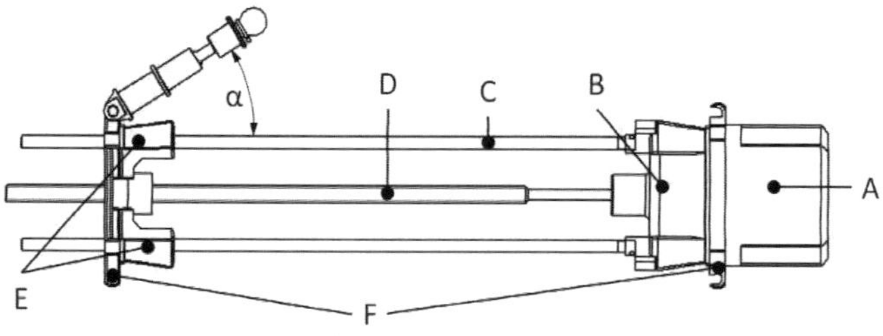

Figure 6. Components of the skeletal structure

6. Spring assembly

The main components of a spring assembly are shown in Figure 7, namely: A: Base body, B: Sliding axle, C: Ball socket, D: Spring stopper, E: Measuring coil, F: Spring.

6.1. Mechanics and Materials

The main task of the spring assembly is to press the track units against the pipe walls resulting in sufficient friction for the drive to move even in adverse conditions. Furthermore, its positional information can be used to determine the thickness of possible sediments.

The base body as well as the spring stopper is made from plastic, so the alternating magnetic field caused by the coil can pass through. Between these two parts a spring is inserted and the travelling distance of the mechanism is 20 mm.

Ball socket, spring stopper and the sliding axle are mounted together, resulting in a movable part, gliding in and lead by the basal body. The latter is flexibly connected to the skeletal structure, ball socket and ball pins of the track units are assembled representing a ball joint.

6.2. Measuring principle

The task is to find a suitable measuring principle that allows determining the length of a spring assembly and thus the radial extension of the tracks. Due to potential corrosion problems, unprotected electrical contacts need to be avoided.

Inductive principles like eddy current sensors seem to be most suitable, because they work in a contactless manner and can be fully encapsulated. According to [14] the equivalent inductance L of an eddy current sensor can be estimated to be

$$L = L_1 - \frac{4\,\pi^2\, f^2\, M^2\, L_2}{R_2^2 + (2\,\pi\, f\, L_2)^2}, \qquad (4)$$

where L_1 is the self-inductance of the sensor, R_2 and L_2 are the equivalent resistance and self-inductance, respectively, of the axle and M is the mutual inductance between the sensor coil and the axle [14]. The displacement and the actual length of a spring assembly can be determined from the measured inductance via its characteristic. A copper core is inserted in a centre bore of the axle, so R_2 decreases, resulting in a higher change in inductance when the axle is displaced.

Assuming the situation shown in Figure 8 with opposite track units (components A and B) staying parallel to the skeletal structure. From the mean lengths of the involved spring units (a, b, c, d) as well as the measured distance between the connecting sheets x and the given dimensions l, q and D the complete outer dimension H can be calculated as

$$H = \frac{\sqrt{2\,(x-l)^2\,(a^2+b^2)-(x-l)^4-(a^2-b^2)^2} + \sqrt{2\,(x-l)^2\,(d^2+c^2)-(x-l)^4-(d^2-c^2)^2}}{2\,(x-l)} + q + D\,. \qquad (5)$$

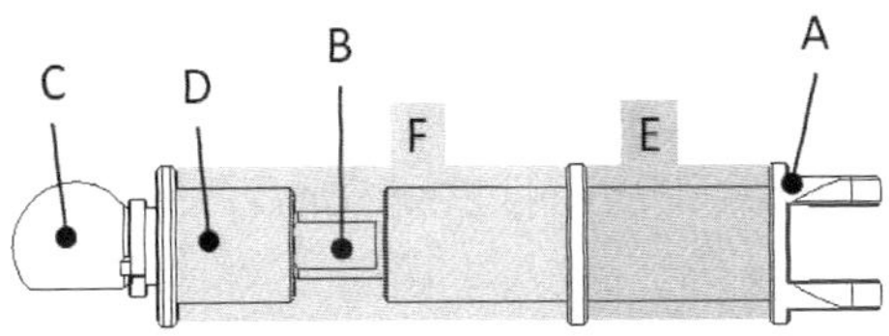

Figure 7. Components of a spring assembly.

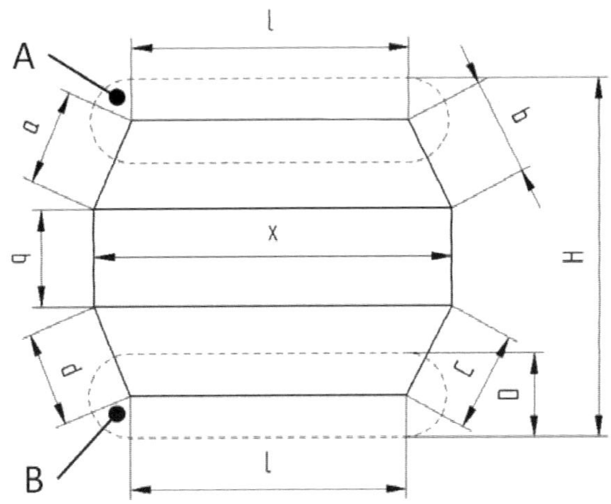

Figure 8. Robot in an arbitrary situation

This is done for each opposite pair of track units, resulting in two values proportional to the thickness of the sediments. To eliminate the impact of pipe deformations a reference measurement in recently cleaned drainages can be done. As a consequence of the 90° angularly displaced tracks used as measure points calcite sediments in between cannot be detected.

7. Conclusion

In this contribution a concept for an in-pipe inspection robot has been presented. The self adapting design includes solutions for passing challenging drainage pipe sections like kinks, junctions and half-open tubes. Additionally, the rather narrow pipe dimensions requested compact systems combining various features. Its behaviour in cases of failure is planned, resulting in a potentially self-disassembling and freewheeling solution. The design also demonstrates measurement strategies for determining the location as well as the thickness of sediments. The materialization and the verification of the presented principles are ongoing.

8. References

[1] Wegmüller M C 2001 *Einflüsse des Bergwassers auf Tiefbau/Tunnelbau* Stäubli AG Zürich ISBN: 3 7266 0052 3

[2] Stur M, Ottner F, Schachinger T and Wriessnig K 2013 Calcium hydroxide (Ca(OH)2) as a component of scaled deposits in tunnel drainage systems *Proceedings of the 11th International Probabilistic Workshop*, pp. 419–438,

[3] Stur M 2011 *Problems of scaling in tunnel drainage systems* Master thesis University of Natural Resources and Life Sciences Vienna

[4] Chen Y, Cui Y, Barrett A G, Chille F and Lassalle F 2019 Investigation of calcite precipitation in the drainage system of railway tunnels *Tunnelling and Underground Space Technology* vol **84** (Elsevier) pp. 45-55

[5] Jung H, Han Y, Chung S, Chun B and Lee Y 2013 Evaluation of advanced drainage treatment for old tunnel drainage system in Korea *Tunnelling and Underground Space Technology* vol **38** (Elsevier) pp. 476-486

[6] Schachinger T, Saliger F, Zagar B, Stur M and Schwab C 2018 Current research by ÖBB Infrastruktur AG on scale monitoring without track closures *Geomechanics and Tunnelling* vol **11** (Wiley) pp. 277-285

[7] Schachinger T, Sperger L, Heissenberger R and Wagner O K 2017 Task Force Drainage (TFD) - The project for life after *Geomechanics and Tunnelling* vol **10** (Wiley) pp. 779-787

[8] Brown L, Carrasco J, Watson S and Lennox B 2019 Elbow Detection in Pipes for Autonomous Navigation of Inspection Robots *Journal of Intelligent & Robotic Systems* vol **95** (Springer) pp. 527-541

[9] Nassiraei A A F, Kawamura Y, Ahrary A, Mikuriya Y and Ishii K 2007 Concept and Design of A Fully Autonomous Sewer Pipe Inspection Mobile Robot "KANTARO" *Proceedings 2007 IEEE International Conference on Robotics and Automation* (IEEE) pp. 136-143

[10] Mills G H, Jackson A E and Richardson R C 2017 Advances in the Inspection of Unpiggable Pipelines *Robotics 2017* (MDPI) pp. 3-8

[11] Shao L, Wang Y and Chen B G X 2015 A Review over State of the Art of In-Pipe Robot *2015 IEEE International Conference on Mechatronics and Automation* (IEEE) pp. 2180-2185

[12] Moghadam M M, Arbabtafti M and Hadi A R 2011 In-pipe inspection crawler adaptable to the pipe interior diameter *International Journal of Robotics and Automation* vol **26** (ACTA Press) pp. 136-145

[13] Nanotec 2015 Brushless DC Motor DF45L024048-A2 *Datasheet*

[14] Tian G Y, Zhao Z X and Baines R W 1998 The research of inhomogeneity in eddy current sensors *Sensors and Actuators* vol **69** (Elsevier) pp. 148-151

Acknowledgment

The authors would like to acknowledge the partial funding of the research presented in this article by the Austrian Research Promotion Agency (FFG) as part of the project 860566 (Drainage-Eye).

IOP Conf. Series: Materials Science and Engineering **831** (2020) 012017 doi:10.1088/1757-899X/831/1/012017

Using Small Displacement Torsor to Simulate the Machining Processes for 3D Tolerance Transfer

Heping Peng [1]* and Zhuoqun Peng [2]

[1] School of Electromechanical & Architectural Engineering, Jianghan University, Wuhan 430056, China

[2] School of Machinery and Automation, Wuhan University of Science and Technology, Wuhan 430081, China

hbjhun_penn@outlook.com

Abstract. This paper proposes a method for 3D tolerance transfer by using small displacement torsor (SDT) to simulate the machining processes. In this method, the machining process of a mechanical part is regarded as a mechanism mainly composed of machine tool, part-holders, machined part, and cutting tools. And the small displacement torsors are employed to describe all defects in the machining process including the variation deviations of the machined surfaces of a part with regards to their nominal positions, the gap deviations associated to each joint between two contact surfaces, etc. Then, the 3D manufacturing tolerancing chains based on the functional tolerances can be established to realize 3D tolerance transfer. The proposed method is applied to a machining part to demonstrate its effectiveness.

1. Introduction

Tolerance is the important product information at each stage of product design, manufacturing, verification, assembly, testing, etc. And it is the concrete reflection of product precision, the result of trade-off between product requirements and development costs [1]. Tolerance transfer, also considered as tolerance analysis and synthesis in manufacturing process planning, is the approach to convert design tolerances into manufacturing tolerances [2]. Jaballi et al. [3] developed 3D tolerance transfer method based on the Technologically and Topologically Related Surface (TTRS). Jiang et al. [4] studied transfer of cylindrical datum in manufacturing process with T-Map model. Villeneuve et al. [5] put forward 3D model to perform manufacturing tolerancing for mechanical parts in which the small displacement torsor (SDT) concept is employed to model the machined parts, part-holders, and machining operations. And Vignat and Villeneuve [6] performed 3D manufacturing tolerancing for turning operation by using the SDT approach. And since then, the SDT approach was further extended to manufacturing tolerance synthesis by Villeneuve and Vignat [7]. In addition, Villeneuve and Vignat [8] proposed the model of manufactured part (MMP) for 3D transfer of tolerances by simulating successive machining processes.

This paper presents a general 3D modeling method for tolerance transfer in the context of manufacturing tolerancing. The proposed method uses small displacement torsors to model all variations in the successive set-ups of the machining process regardless of the causes of these variations. Then, the influence of the process planning on the respect of functional tolerances is modelled as 3D tolerances chain to explore the manufacturing tolerancing transfer. Finally, application to a machining part is introduced to verify feasibility of the proposed method.

2. Small displacement torsor

Generally, any variation of a geometrical feature from the nominal position can be represented by using the SDT with three rotation components (α, β, γ) and three translation components (u, v, w) [9]. SDT $\{\boldsymbol{\tau}_{P_i/P}\}_{(O_i,R_i)}$ synthesizes the position and orientation of an associated feature P_i relative to its nominal feature at a given point O_i in a local reference frame $\{R_i\}$ and consists of original vector $\boldsymbol{\varphi}$ and dual vector $\boldsymbol{\varepsilon}_{O_i}$ [10].

$$\{\boldsymbol{\tau}_{P_i/P}\}_{(O_i,R_i)} = \{\boldsymbol{\varphi} \quad \boldsymbol{\varepsilon}_{O_i}\} = \begin{Bmatrix} \alpha & u \\ \beta & v \\ \gamma & w \end{Bmatrix}_{(O_i,R_i)} \tag{1}$$

For example, the torsor represented the relative positions of the associated plane P_i with regard to its nominal plane at a given point O_i in the local reference frame $\{R_i\}$ with axis z_i is the normal of the plane can be expressed as:

$$\{\boldsymbol{\tau}_{P_i/P}\}_{(O_i,R_i)} = \{\boldsymbol{\varphi} \quad \boldsymbol{\varepsilon}_{O_i}\} = \begin{Bmatrix} \alpha & U \\ \beta & U \\ U & w \end{Bmatrix}_{(O_i,R_i)} \tag{2}$$

where capital U represents the undetermined component in the expression of a torsor.
The two following properties will be defined to calculate operations on these torsors [5]:

$$\forall a \in R,\ a + U = U \tag{3}$$

$$\forall a,b \in R^2,\ a \cdot U + b \cdot U = U \tag{4}$$

Suppose O is a given point of the global reference frame $\{R_0\}$, the expression of relative translation vector ε_O at point O can be derived by using a linearization of the relative rotations, from which, one obtains the transfer expression of the torsor:

$$\{\boldsymbol{\tau}_{P_i/P}\}_{(O,R_0)} = \left\{\mathbf{R}_{0,i} \cdot \boldsymbol{\varphi} \quad \mathbf{R}_{0,i} \cdot (\boldsymbol{\varepsilon}_{O_i} + (\mathbf{R}_{0,i}^T \cdot \overrightarrow{OO_i}) \times \boldsymbol{\varphi})\right\} \tag{5}$$

where $\mathbf{R}_{0,i}$ is the rotation matrix from $\{R_0\}$ to $\{R_i\}$, $\mathbf{R}_{0,i}^T$ is the transposed matrix of $\mathbf{R}_{0,i}$, and $\overrightarrow{OO_i}$ is the translation vector from $\{R_0\}$ to $\{R_i\}$ expressed in $\{R_0\}$.

3. Modelling of variations in a machining process

In production practice, the process engineers make process plans by selecting the appropriate machining processes and production equipment to ensure the functional requirement of machined parts. Considering that a process plan consists of several different set-ups, for each set-up there are several SDTs need to be defined as follows:

- *For machined part*

$\boldsymbol{\tau}_{P/R}^{S_k}$: global SDT characterizes the positioning variations of the machined part with respect to its nominal position in set-up S_k.

$\boldsymbol{\tau}_{P_i/R}^{S_k}$: deviation torsor of the machined surface P_i with regards to its nominal position in set-up S_k (variations of the machining operation).

$\boldsymbol{\tau}_{P_i/P}$: deviation torsor expresses the orientation and position variations of surface P_i relatively to its nominal position on part P. Suppose surface P_i is manufactured in set-up S_k, this deviation torsor can be calculated as:

$$\boldsymbol{\tau}_{P_i/P} = \boldsymbol{\tau}_{P_i/R}^{S_k} - \boldsymbol{\tau}_{P/R}^{S_k} \tag{6}$$

- *For part-holder*

$\boldsymbol{\tau}_{H/R}^{S_k}$: global SDT of part-holder H relatively to its nominal position in set-up S_k.

$\tau^{S_k}_{H_i/H}$: deviation torsor expresses the geometric variations of surface H_i relatively to its nominal position on part-holder in set-up k.

- *For machining operation*

$\tau^{S_k}_{M/R}$: torsor characterizes variations of a machine tool relatively to its nominal position in set-up S_k.

$\tau^{S_k}_{M_i/M}$: deviation torsor expresses the geometric variations of surface M_i relatively to its nominal position on the machining operation M in set-up S_k.

- *For interface of part and part-holder*

$\tau^{S_k}_{P_i/H_i}$: gap torsor that expresses the variations of the interface between surface P_i of the machined part and thc corresponding surface H_i of the part-holder in set-up k. Given that the parts do not interpenetrate at the contacts, each fixed component of the torsors is regarded as nil.

The main link types of the interfaces between the machined surfaces and the positioning surfaces in manufacturing operations and corresponding torsors are list in table 1. The undetermined components in the link torsors are denoted by "U", which show the components that leave the characteristic invariance of corresponding direction.

Table 1. Main link types and the corresponding torsors.

Link types	Coordinate system	Link torsors	Link types	Coordinate system	Link torsors
Planar	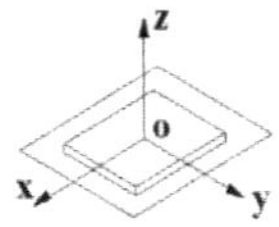	$\begin{Bmatrix} \alpha & U \\ \beta & U \\ U & w \end{Bmatrix}$	Edge Slider	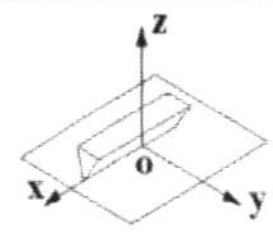	$\begin{Bmatrix} U & U \\ \beta & U \\ U & w \end{Bmatrix}$
Prismatic	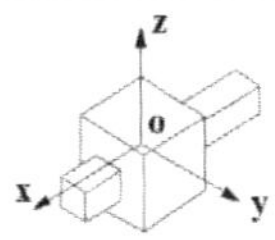	$\begin{Bmatrix} \alpha & U \\ \beta & v \\ \gamma & w \end{Bmatrix}$	Point Slider	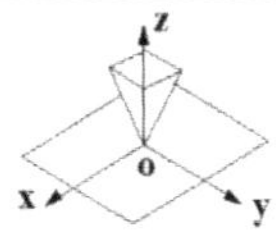	$\begin{Bmatrix} U & U \\ U & U \\ U & w \end{Bmatrix}$
Cylindrical	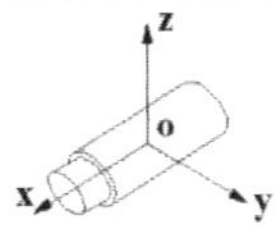	$\begin{Bmatrix} U & U \\ \beta & v \\ \gamma & w \end{Bmatrix}$	Cylindrical Slider	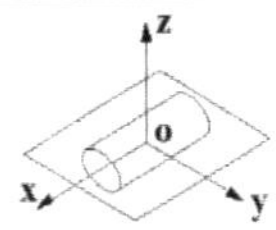	$\begin{Bmatrix} U & U \\ \beta & U \\ U & w \end{Bmatrix}$
Spherical	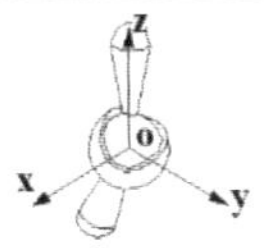	$\begin{Bmatrix} U & u \\ U & v \\ U & w \end{Bmatrix}$	Spherical Slider	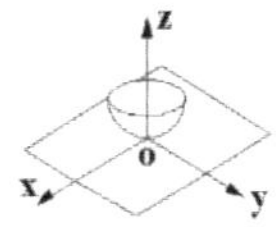	$\begin{Bmatrix} U & U \\ U & U \\ U & w \end{Bmatrix}$

The developed manufacturing model regards each manufacturing set-up as a mechanism consisting of a machine tool, several manufacturing operations, a machined part and its surface, and a part-holder and its surface. As shown in figure 1, for any set of two interacting surfaces (Pi, Hi), we have:

$$\tau_{P_i/H_i} = \tau_{P_i/P} + \tau_{P/R} - \tau_{H/R} - \tau_{H_i/H} \quad (7)$$

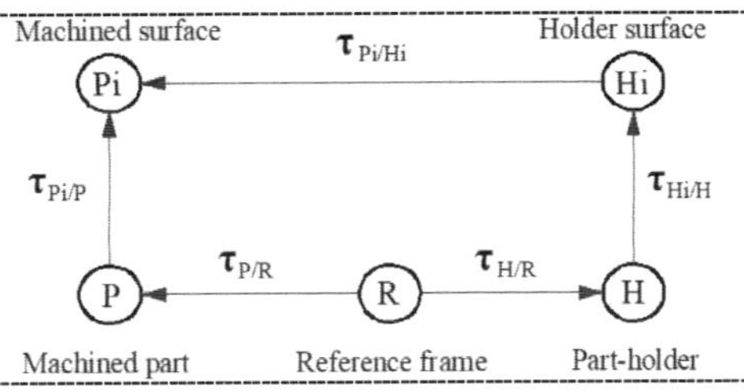

Figure 1. Various torsors of contact surfaces.

4. Three-dimensional tolerance transfer

Tolerance transfer refers to as tolerance analysis and synthesis in process planning. The purpose of tolerance analysis is to accumulate the effects of various variations in machining operation to determine if each functional tolerance of the part will be respected. Tolerance synthesis is to obtain more loose machining tolerances in each set-up while respecting the functional tolerances of the part.

Torsor $\boldsymbol{\tau}_{P_b/P_a}$ can represent the functional tolerance between two machined surfaces P_a and P_b of part P. It can be expressed by the equation:

$$\boldsymbol{\tau}_{P_b/P_a} = \boldsymbol{\tau}_{P_b/P} + \boldsymbol{\tau}_{P/P_a} = \boldsymbol{\tau}_{P_b/P} - \boldsymbol{\tau}_{P_a/P} \tag{8}$$

Suppose surfaces P_a and P_b are machined in set-ups S_1 and S_2 respectively, according to equations (6) and (7), equation (8) becomes:

$$\boldsymbol{\tau}_{P_b/P_a} = (\boldsymbol{\tau}^{S_2}_{P_b/R} + \boldsymbol{\tau}_{P_i/P} - \boldsymbol{\tau}^{S_2}_{P_i/H_i} - \boldsymbol{\tau}^{S_2}_{H_i/H} - \boldsymbol{\tau}^{S_2}_{H/R}) - (\boldsymbol{\tau}^{S_1}_{P_a/R} + \boldsymbol{\tau}_{P_i/P} - \boldsymbol{\tau}^{S_1}_{P_i/H_i} - \boldsymbol{\tau}^{S_1}_{H_i/H} - \boldsymbol{\tau}^{S_1}_{H/R}) \tag{9}$$

Thus, the relationship between functional tolerances and various variations in each process planning can be established. This relationship can guide process engineers to perform manufacturing tolerance analysis and synthesis.

5. Illustrative example

A mechanical part shown in figure 2 is used to demonstrate the proposed method. The functional requirements being studied are a location and a parallelism of plane P_7 with respect to datum **A** on plane P_1. Figures 3-5 show the machining process of this part, which consists of three set-ups performed on a numerical control (NC) machine-tool.

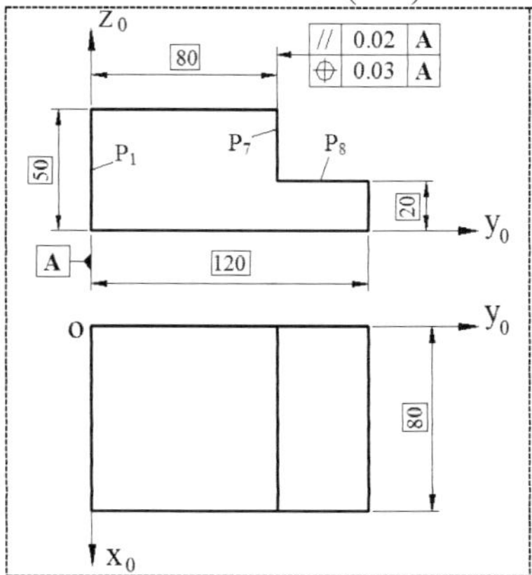

Figure 2. Machined part geometry.

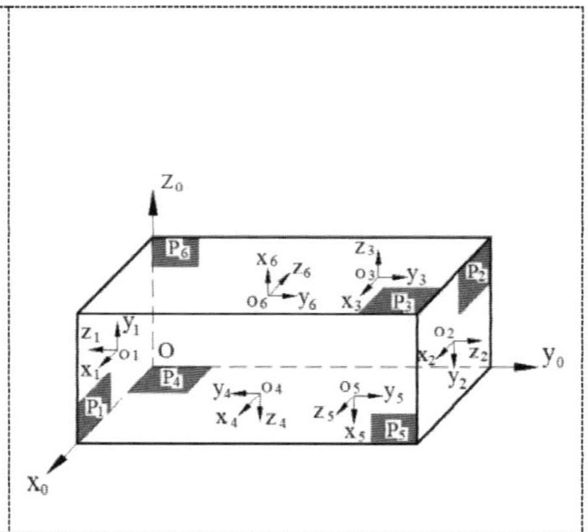

Figure 3. Primarily machined part and reference frames.

5.1 Set-up 10

As shown in figure 3, the machined surfaces are marked as P_i, the local reference frame R_i (O_i, x_i, y_i, z_i) for each machined surface is defined as: Axis z_i is normal to surface P_i pointing towards the outside of the entity; Origin O_i, axes x_i and y_i of the reference frame belong to surface P_i.

In this set-up, all surfaces are primarily machined, that is, there are no positioning surfaces. In the global reference frame R_0 (O, x_0, y_0, z_0) of figure 3, we consider: $\{\boldsymbol{\tau}_{P/R}\}_{(O,R_0)} = \{0 \quad 0\}$. And:

$$\boldsymbol{\tau}_{P_i/P} = \boldsymbol{\tau}_{P_i/R} - \boldsymbol{\tau}_{P/R} = \boldsymbol{\tau}_{P_i/R} \tag{10}$$

So we have:

$$\{\boldsymbol{\tau}_{P_i/P}\}_{(O_i,R_i)} = \begin{Bmatrix} \alpha_{P_i} & U_P \\ \beta_{P_i} & U_P \\ U_P & w_{P_i} \end{Bmatrix}, \quad i \in \{1, 2, \cdots 6\} \tag{11}$$

5.2 Set-up 20

In this set-up, surface 1 of the part is machined. The machined part is positioned on a plane (main positioning surface H_4/P_4), a cylinder (2nd positioning surface H_2/P_2) and a sphere (3rd positioning surface H_6/P_6) in an isostatic set-up as shown in figure 4. Here, the part-holder support points are also marked as O_i in a local frame R_i (O_i, x_i, y_i, z_i), $i\in\{2, 4, 6\}$.

The global SDT $\boldsymbol{\tau}^{20}_{P/R}$ of the machined part can be obtained by combining the torsors associated with joints between the part and the part-holder. In a local frame R_i (O_i, x_i, y_i, z_i), $i\in\{2, 4, 6\}$, suppose the global variations of part-holder H are integrated within the deviation torsor of surface H_4 relative to its nominal position in this set up for the main positioning surface H_4, and combine with the expression of the torsor matrices in table 1, we have:

$$\{\boldsymbol{\tau}^{20}_{H_4/H}\}_{(O_4,R_4)}=\begin{Bmatrix}\alpha^{20}_{H_4} & U_H\\ \beta^{20}_{H_4} & U_H\\ U_H & w^{20}_{H_4}\end{Bmatrix},\ \{\boldsymbol{\tau}^{20}_{H_2/H}\}_{(O_2,R_2)}=\begin{Bmatrix}U_H & U_H\text{-}r_c\beta^{20}_{H_2}\\ \beta^{20}_{H_2} & v^{20}_{H_2}+r_cU_H\\ \gamma^{20}_{H_2} & w^{20}_{H_2}\end{Bmatrix}\ \text{and}\ \{\boldsymbol{\tau}^{20}_{H_6/H}\}_{(O_6,R_6)}=\begin{Bmatrix}U_H & u^{20}_{H_6}\text{-}r_sU_H\\ U_H & v^{20}_{H_6}+r_sU_H\\ U_H & w^{20}_{H_6}\end{Bmatrix} \quad (12)$$

where the 2nd positioning surface H_2 is a cylindrical surface with radius r_c, the 3rd positioning surface H_6 is a spherical surface whose radius is r_s.

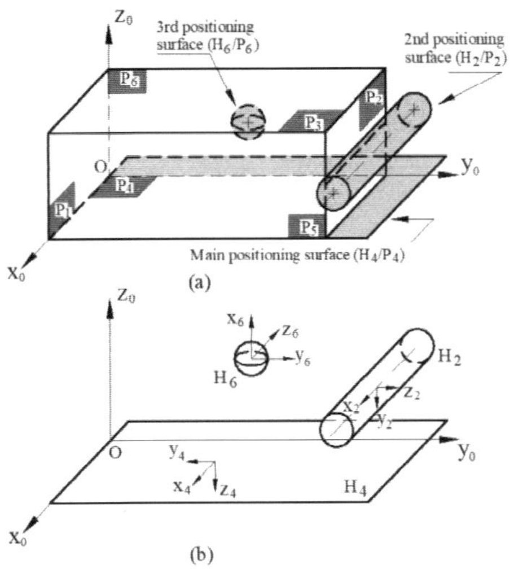

Figure 4. Set-up20: Milling of surface P_1.

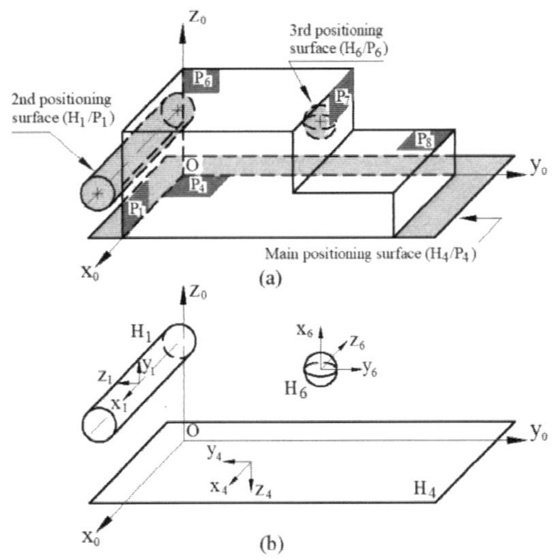

Figure 5. Set-up30: Milling of surface P_7.

Then we can calculate the torsors $\{\boldsymbol{\tau}^{20}_{P_i/P}\}_{(O,R_0)}$ and $\{\boldsymbol{\tau}^{20}_{H_i/H}\}_{(O,R_0)}$ $i\in\{2,4,6\}$ at point O expressed in the global reference frame R_0 (O, x_0, y_0, z_0) by using equation (5), suppose the origin coordinate of the local reference frame R_i is O_i (a_i, b_i, c_i), $i\in\{2,4,6\}$ (see figure 3), the rotation matrices $\mathbf{R}_{0,i}$ from $\{R_0\}$ to $\{R_i\}$ respectively are:

$$\mathbf{R}_{0,4}=\begin{Bmatrix}1 & 0 & 0\\ 0 & -1 & 0\\ 0 & 0 & -1\end{Bmatrix},\ \mathbf{R}_{0,2}=\begin{Bmatrix}1 & 0 & 0\\ 0 & 0 & 1\\ 0 & -1 & 0\end{Bmatrix},\ \mathbf{R}_{0,6}=\begin{Bmatrix}0 & 0 & -1\\ 0 & 1 & 0\\ 1 & 0 & 0\end{Bmatrix} \quad (13)$$

Then we can get the expressions of $\{\boldsymbol{\tau}_{P_i/P}\}_{(O,R_0)}$ and $\{\boldsymbol{\tau}^{20}_{H_i/H}\}_{(O,R_0)}$, $i\in\{2,4,6\}$. And according to equation (7), we can get the gap SDT $\{\boldsymbol{\tau}^{20}_{P_i/H_i}\}_{(O,R_0)}$ at point O expressed in global reference frame R_0 (O, x_0, y_0, z_0).

$$\{\boldsymbol{\tau}^{20}_{P_4/H_4}\}_{(O,R_0)}=\begin{Bmatrix} \alpha_{P_4}+\alpha^{20}_{P}-\alpha^{20}_{H}-\alpha^{20}_{H_4} & U_P-b_4U_P+c_4\beta_{P_4}+u^{20}_{P}-u^{20}_{H}-U_H+b_4U_H-c_4\beta^{20}_{H_4} \\ -\beta_{P_4}+\beta^{20}_{P}-\beta^{20}_{H}+\beta^{20}_{H_4} & -U_P+a_4U_P+c_4\alpha_{P_4}+v^{20}_{P}-v^{20}_{H}+U_H-a_4U_H-c_4\alpha^{20}_{H_4} \\ -U_P+\gamma^{20}_{P}-\gamma^{20}_{H}+U_H & -w_{P_4}-a_4\beta_{P_4}-b_4\alpha_{P_4}+w^{20}_{P}-w^{20}_{H}+w^{20}_{H_4}+a_4\beta^{20}_{H_4}+b_4\alpha^{20}_{H_4} \end{Bmatrix} \tag{14}$$

$$\{\boldsymbol{\tau}^{20}_{P_2/H_2}\}_{(O,R_0)}=\begin{Bmatrix} \alpha_{P_2}+\alpha^{20}_{P}-\alpha^{20}_{H}-U_H & U_P-c_2U_P-b_2\beta_{P_2}+u^{20}_{P}-u^{20}_{H}-U_H+r_c\beta^{20}_{H_2}+c_2\gamma^{20}_{H_2}+b_2\beta^{20}_{H_2} \\ U_P+\beta^{20}_{P}-\beta^{20}_{H}-\gamma^{20}_{H_2} & w_{P_2}+a_2\beta_{P_2}+c_2\alpha_{P_2}+v^{20}_{P}-v^{20}_{H}-w^{20}_{H_2}-a_2\beta^{20}_{H_2}-c_2U_H \\ -\beta_{P_2}+\gamma^{20}_{P}-\gamma^{20}_{H}+\beta^{20}_{H_2} & -U_P+a_2U_P-b_2\alpha_{P_2}+w^{20}_{P}-w^{20}_{H}+v^{20}_{H_2}+r_cU_H-a_2\gamma^{20}_{H_2}+b_2U_H \end{Bmatrix} \tag{15}$$

$$\{\boldsymbol{\tau}^{20}_{P_6/H_6}\}_{(O,R_0)}=\begin{Bmatrix} -U_P+\alpha^{20}_{P}-\alpha^{20}_{H}+U_H & -w_{P_6}-c_6\beta_{P_6}+b_6\alpha_{P_6}+u^{20}_{P}-u^{20}_{H}+w^{20}_{H_6}-b_6U_H+c_6U_H \\ \beta_{P_6}+\beta^{20}_{P}-\beta^{20}_{H}-U_H & U_P-c_6U_P-a_6\alpha_{P_6}+v^{20}_{P}-v^{20}_{H}-v^{20}_{H_6}-r_sU_H+c_6U_H+a_6U_H \\ \alpha_{P_6}+\gamma^{20}_{P}-\gamma^{20}_{H}-U_H & U_P+b_6U_P+a_6\beta_{P_6}+w^{20}_{P}-w^{20}_{H}-u^{20}_{H_6}+r_sU_H-b_6U_H-a_6U_H \end{Bmatrix} \tag{16}$$

Suppose $\{\boldsymbol{\tau}_{P_i/H_i}\}_{(O,R_0)}=\{\boldsymbol{\varphi}\quad \boldsymbol{\varepsilon}_O\}$ then,

$$\{\boldsymbol{\tau}_{P_i/H_i}\}_{(O_i,R_i)}=\left\{R^T_{0,i}\cdot\boldsymbol{\varphi}\quad R^T_{0,i}\cdot(\varepsilon_O+\boldsymbol{\varphi}\times\overrightarrow{OO_i})\right\} \tag{17}$$

Using equation (17) to calculate the gap SDT $\{\boldsymbol{\tau}^{20}_{P_i/H_i}\}_{(O_i,R_i)}$ at point O_i expressed in local reference frame R_i (O_i, x_i, y_i, z_i) and considering the properties (3) and (4), one gets:

$$\{\boldsymbol{\tau}^{20}_{P_4/H_4}\}_{(O_4,R_4)}=\begin{Bmatrix} \alpha_{P_4}+\alpha^{20}_{P}-\alpha^{20}_{H}-\alpha^{20}_{H_4} & U \\ \beta_{P_4}-\beta^{20}_{P}+\beta^{20}_{H}-\beta^{20}_{H_4} & U \\ U & w_{P_4}-w^{20}_{P}+w^{20}_{H}-w^{20}_{H_4}-b_4\alpha^{20}_{P}+b_4\alpha^{20}_{H}+a_4\beta^{20}_{P}-a_4\beta^{20}_{H} \end{Bmatrix} \tag{18}$$

$$\{\boldsymbol{\tau}^{20}_{P_2/H_2}\}_{(O_2,R_2)}=\begin{Bmatrix} U & U \\ \beta_{P_2}-\gamma^{20}_{P}+\gamma^{20}_{H}-\beta^{20}_{H_2} & U \\ U & w_{P_2}+v^{20}_{P}-v^{20}_{H}-w^{20}_{H_2}-c_2\alpha^{20}_{P}+c_2\alpha^{20}_{H}+a_2\gamma^{20}_{P}-a_2\gamma^{20}_{H} \end{Bmatrix} \tag{19}$$

$$\{\boldsymbol{\tau}^{20}_{P_6/H_6}\}_{(O_6,R_6)}=\begin{Bmatrix} U & U \\ U & U \\ U & w_{P_6}-u^{20}_{P}+u^{20}_{H}-w^{20}_{H_6}-c_6\beta^{20}_{P}+c_6\beta^{20}_{H}+b_6\gamma^{20}_{P}-b_6\gamma^{20}_{H} \end{Bmatrix} \tag{20}$$

Considering the hierarchy of the part/part-holder positioning in an isostatic set-up, the components of $\{\boldsymbol{\tau}^{20}_{P_4/H_4}\}_{(O_4,R_4)}$ are nil because the contact between the two main positioning surfaces (H_4/P_4) has no interpenetrating parts. One gets the following equations:

$$\begin{cases} \alpha_{P_4}+\alpha^{20}_{P}-\alpha^{20}_{H}-\alpha^{20}_{H_4}=0 \\ \beta_{P_4}-\beta^{20}_{P}+\beta^{20}_{H}-\beta^{20}_{H_4}=0 \\ w_{P_4}-w^{20}_{P}+w^{20}_{H}-w^{20}_{H_4}-b_4\alpha^{20}_{P}+b_4\alpha^{20}_{H}+a_4\beta^{20}_{P}-a_4\beta^{20}_{H}=0 \end{cases} \tag{21}$$

Similarly, for the second and third positioning surfaces, we have:

$$\begin{cases} \beta_{P_2}-\gamma^{20}_{P}+\gamma^{20}_{H}-\beta^{20}_{H_2}=0 \\ w_{P_2}+v^{20}_{P}-v^{20}_{H}-w^{20}_{H_2}-c_2\alpha^{20}_{P}+c_2\alpha^{20}_{H}+a_2\gamma^{20}_{P}-a_2\gamma^{20}_{H}=0 \end{cases} \tag{22}$$

$$w_{P_6}-u^{20}_{P}+u^{20}_{H}-w^{20}_{H_6}-c_6\beta^{20}_{P}+c_6\beta^{20}_{H}+b_6\gamma^{20}_{P}-b_6\gamma^{20}_{H}=0 \tag{23}$$

Then, $\{\boldsymbol{\tau}^{20}_{P/R}\}_{(O,R_0)}$ can be derived as:

$$\{\boldsymbol{\tau}^{20}_{P/R}\}_{(O,R_0)}=\begin{Bmatrix} -\alpha_{P_4}+\alpha^{20}_{H}+\alpha^{20}_{H_4} & w_{P_6}+u^{20}_{H}-w^{20}_{H_6}-c_6\beta_{P_4}+c_6\beta^{20}_{H_4}+b_6\beta_{P_2}-b_6\beta^{20}_{H_2} \\ \beta_{P_4}+\beta^{20}_{H}-\beta^{20}_{H_4} & -w_{P_2}+v^{20}_{H}+w^{20}_{H_2}-c_2\alpha_{P_4}+c_2\alpha^{20}_{H_4}-a_2\beta_{P_2}+a_2\beta^{20}_{H_2} \\ \beta_{P_2}+\gamma^{20}_{H}-\beta^{20}_{H_2} & w_{P_4}+w^{20}_{H}-w^{20}_{H_4}+b_4\alpha_{P_4}-b_4\alpha^{20}_{H_4}+a_4\beta_{P_4}-a_4\beta^{20}_{H_4} \end{Bmatrix} \tag{24}$$

Using equation (17), we get:

$$\{\boldsymbol{\tau}^{20}_{P/R}\}_{(O_1,R_1)}=\begin{Bmatrix} -\alpha_{P_4}+\alpha^{20}_{H}+\alpha^{20}_{H_4} & w_{P_6}+u^{20}_{H}-w^{20}_{H_6}-c_6\beta_{P_4}+c_6\beta^{20}_{H_4}+b_6\beta_{P_2}-b_6\beta^{20}_{H_2}+c_1\beta_{P_4}+c_1\beta^{20}_{H}-c_1\beta^{20}_{H_4}-b_1\beta_{P_2}-b_1\gamma^{20}_{H}+b_1\beta^{20}_{H_2} \\ \beta_{P_2}+\gamma^{20}_{H}-\beta^{20}_{H_2} & w_{P_4}+w^{20}_{H}-w^{20}_{H_4}+b_4\alpha_{P_4}-b_4\alpha^{20}_{H_4}+a_4\beta_{P_4}-a_4\beta^{20}_{H_4}-b_1\alpha_{P_4}+b_1\alpha^{20}_{H}+b_1\alpha^{20}_{H_4}-a_1\beta_{P_4}-a_1\beta^{20}_{H}+a_1\beta^{20}_{H_4} \\ -\beta_{P_4}-\beta^{20}_{H}+\beta^{20}_{H_4} & w_{P_2}-v^{20}_{H}-w^{20}_{H_2}+c_2\alpha_{P_4}-c_2\alpha^{20}_{H_4}+a_2\beta_{P_2}-a_2\beta^{20}_{H_2}-c_1\alpha_{P_4}+c_1\alpha^{20}_{H}+c_1\alpha^{20}_{H_4}-a_1\beta_{P_2}-a_1\gamma^{20}_{H}+a_1\beta^{20}_{H_2} \end{Bmatrix} \tag{25}$$

The torsor describing the defects due to the machining operation with regards to the machine is:

$$\{\boldsymbol{\tau}^{20}_{P_1/R}\}_{(O_1,R_1)} = \begin{Bmatrix} \alpha_{P_1} & U_P \\ \beta_{P_1} & U_P \\ U_P & w_{P_1} \end{Bmatrix} \tag{26}$$

And

$$\{\boldsymbol{\tau}_{P_1/P}\}_{(O_1,R_1)} = \{\boldsymbol{\tau}^{20}_{P_1/R}\}_{(O_1,R_1)} - \{\boldsymbol{\tau}^{20}_{P/R}\}_{(O_1,R_1)} \tag{27}$$

Then:

$$\{\boldsymbol{\tau}_{P_1/P}\}_{(O_1,R_1)} = \begin{Bmatrix} \alpha_{P_1} + \alpha_{P_4} - \alpha^{20}_{H} - \alpha^{20}_{H_4} & U \\ \beta_{P_1} - \beta_{P_2} - \gamma^{20}_{H} + \beta^{20}_{H_2} & U \\ U & w_{P_1} - w_{P_2} + v^{20}_{H} + w^{20}_{H_2} - c_2\alpha_{P_4} + c_2\alpha^{20}_{H_4} - a_2\beta_{P_2} + a_2\beta^{20}_{H_2} + c_1\alpha_{P_4} - c_1\alpha^{20}_{H} - c_1\alpha^{20}_{H_4} + a_1\beta_{P_2} + a_1\gamma^{20}_{H} - a_1\beta^{20}_{H_2} \end{Bmatrix} \tag{28}$$

According to equation (5), for the machined surface P_1 in this set-up we have:

$$\{\boldsymbol{\tau}_{P_1/P}\}_{(O,R_0)} = \begin{Bmatrix} \alpha_{P_1} + \alpha_{P_4} - \alpha^{20}_{H} - \alpha^{20}_{H_4} & U \\ U & -w_{P_1} + w_{P_2} - v^{20}_{H} - w^{20}_{H_2} + c_2\alpha_{P_4} - c_2\alpha^{20}_{H_4} + a_2\beta_{P_2} - a_2\beta^{20}_{H_2} - a_1\beta_{P_1} + c_1\alpha_{P_1} \\ \beta_{P_1} - \beta_{P_2} - \gamma^{20}_{H} + \beta^{20}_{H_2} & U \end{Bmatrix} \tag{29}$$

5.3 Set-up 30

In this set-up, surfaces 7 and 8 are machined. As shown in figure 5, the part is positioned on a plane (main positioning surface H_4/P_4), a cylinder (2nd positioning surface H_1/P_1) and a sphere (3rd positioning surface H_6/P_6) in an isostatic set-up. For each positioning device H_i of the part-holder in a local frame R_i (O_i, x_i, y_i, z_i), $i \in \{1, 4, 6\}$.

Consider the main positioning surface H4, and suppose the global variations of part-holder H are integrated within the deviation torsor of surface H_4 relative to its nominal position in set up 30, we have:

$$\{\boldsymbol{\tau}^{30}_{H_4/H}\}_{(O_4,R_4)} = \begin{Bmatrix} \alpha^{30}_{H_4} & U_H \\ \beta^{30}_{H_4} & U_H \\ U_H & w^{30}_{H_4} \end{Bmatrix}, \quad \{\boldsymbol{\tau}^{30}_{H_1/H}\}_{(O_1,R_1)} = \begin{Bmatrix} U_H & U_H - r_c\beta^{30}_{H_1} \\ \beta^{30}_{H_1} & v^{30}_{H_1} + r_c U_H \\ \gamma^{30}_{H_1} & w^{30}_{H_1} \end{Bmatrix}, \quad \{\boldsymbol{\tau}^{30}_{H_6/H}\}_{(O_6,R_6)} = \begin{Bmatrix} U_H & u^{30}_{H_6} - r_s U_H \\ U_H & v^{30}_{H_6} + r_s U_H \\ U_H & w^{30}_{H_6} \end{Bmatrix} \tag{30}$$

Suppose the origin coordinate of the local reference frame R_i is O_i (a_i, b_i, c_i), $i \in \{1, 4, 6\}$ (see figure 3), we can calculate the torsors $\{\boldsymbol{\tau}_{P_i/P}\}_{(O,R_0)}$ and $\{\boldsymbol{\tau}_{H_i/H}\}_{(O,R_0)}$ at point O expressed in the global reference frame R_0 (O, x_0, y_0, z_0) by using equation (5). According to equation (7), we can get the gap SDT $\{\boldsymbol{\tau}^{30}_{P_i/H_i}\}_{(O,R_0)}$ at point O expressed in global reference frame R_0 (O, x_0, y_0, z_0). Using equation (17) to calculate the gap SDT $\{\boldsymbol{\tau}^{30}_{P_i/H_i}\}_{(O_i,R_i)}$ at point O_i expressed in reference frame R_i (O_i, x_i, y_i, z_i). So, $\{\boldsymbol{\tau}^{30}_{P/R}\}_{(O,R_0)}$ can be derived as:

$$\{\boldsymbol{\tau}^{30}_{P/R}\}_{(O,R_0)} = \begin{Bmatrix} -\alpha_{P_4} + \alpha^{30}_{H} + \alpha^{20}_{H_4} & w_{P_6} + u^{30}_{H} - w^{30}_{H_6} - c_6\beta_{P_4} + c_6\beta^{20}_{H_4} - b_6\beta_{P_1} + b_6\beta_{P_2} + b_6\gamma^{20}_{H} - b_6\beta^{20}_{H_2} + b_6\beta^{30}_{H_1} \\ \beta_{P_4} + \beta^{30}_{H} - \beta^{20}_{H_4} & w_{P_1} - w_{P_2} + v^{20}_{H} + w^{20}_{H_2} - c_2\alpha_{P_4} + c_2\alpha^{20}_{H_4} - a_2\beta_{P_2} + a_2\beta^{20}_{H_2} + v^{30}_{H} - w^{30}_{H_1} - c_1\alpha^{20}_{H} + a_1\beta_{P_1} - a_1\beta^{30}_{H_1} \\ -\beta_{P_1} + \beta_{P_2} + \gamma^{20}_{H} - \beta^{20}_{H_2} + \gamma^{30}_{H} + \beta^{30}_{H_1} & w_{P_4} + w^{30}_{H} - w^{30}_{H_4} + b_4\alpha_{P_4} - b_4\alpha^{20}_{H_4} + a_4\beta_{P_4} - a_4\beta^{20}_{H_4} \end{Bmatrix} \tag{31}$$

And $\{\boldsymbol{\tau}_{P_7/P}\}_{(O,R_0)} = \boldsymbol{\tau}^{30}_{P_7/R} - \boldsymbol{\tau}^{30}_{P/R}$, according to equation (5), we have:

$$\{\boldsymbol{\tau}^{30}_{P_7/R}\}_{(O,R_0)} = \begin{Bmatrix} \alpha^{30}_{P_7} & U_P - c_7U_P - b_7\beta^{30}_{P_7} \\ U_P & w^{30}_{P_7} + a_7\beta^{30}_{P_7} + c_7\alpha^{30}_{P_7} \\ -\beta^{30}_{P_7} & -U_P + a_7U_P - b_7\alpha^{30}_{P_7} \end{Bmatrix} \tag{32}$$

For the machined surface P_7 in set-up 30 one gets:

$$\{\tau_{P_7/P}\}_{(O,R_0)} = \begin{Bmatrix} \alpha^{30}_{P_7} + \alpha_{P_4} - \alpha^{30}_{H} - \alpha^{20}_{H_4} & U_P - c_7U_P - b_7\beta^{30}_{P_7} - w_{P_6} - u^{30}_{H} + w^{30}_{H_6} + c_6\beta_{P_4} - c_6\beta^{20}_{H_4} + b_6\beta_{P_1} - b_6\beta_{P_2} - b_6\gamma^{20}_{H} + b_6\beta^{20}_{H_2} - b_6\beta^{30}_{H_1} \\ U_P - \beta_{P_4} - \beta^{30}_{H} + \beta^{20}_{H_4} & w^{30}_{P_7} + a_7\beta^{30}_{P_7} + c_7\alpha^{30}_{P_7} - w_{P_1} + w_{P_2} - v^{20}_{H} - w^{20}_{H_2} + c_2\alpha_{P_4} - c_2\alpha^{20}_{H_4} + a_2\beta_{P_2} - a_2\beta^{20}_{H_2} - v^{30}_{H} + w^{30}_{H_1} + c_1\alpha^{20}_{H} - a_1\beta_{P_1} + a_1\beta^{30}_{H_1} \\ -\beta^{30}_{P_7} + \beta_{P_1} - \beta_{P_2} - \gamma^{20}_{H} + \beta^{20}_{H_2} - \gamma^{30}_{H} - \beta^{30}_{H_1} & -U_P + a_7U_P - b_7\alpha^{30}_{P_7} - w_{P_4} - w^{30}_{H} + w^{30}_{H_4} - b_4\alpha_{P_4} + b_4\alpha^{20}_{H_4} - a_4\beta_{P_4} + a_4\beta^{20}_{H_4} \end{Bmatrix} \tag{33}$$

5.4 Respect of the functional tolerances

In this section, we establish the relations between the functional tolerances and the variations in a machining process to verify the validity of the manufacturing process.

• Consider the parallelism tolerance requirement between surface P_1 and P_7. The variation torsor of surface P_7 relative to surface P_1 (datum surface) is:

$$\{\tau_{P_7/P_1}\}_{(O,R_0)} = \begin{Bmatrix} \alpha_{P_7}^{30} - \alpha_{H}^{30} - \alpha_{P_1} + \alpha_{H}^{20} & -c_7 U_P - b_7 \beta_{P_7}^{30} - u_{H}^{30} + w_{H_6}^{30} + b_6 \beta_{P_1} - b_6 \gamma_{H}^{20} - b_6 \beta_{H_1}^{30} + u_{H}^{20} - w_{H_6}^{20} - c_1 U_P - b_1 \beta_{P_1} \\ U_P - \beta_{H}^{30} + U_P + \beta_{H}^{20} & w_{P_7}^{30} + a_7 \beta_{P_7}^{30} + c_7 \alpha_{P_7}^{30} - v_{H}^{30} + w_{H_1}^{30} + c_1 \alpha_{H}^{20} + a_1 \beta_{H_1}^{30} - c_1 \alpha_{P_1} \\ -\beta_{P_7}^{30} - \gamma_{H}^{30} - \beta_{H_1}^{30} & -U_P + a_7 U_P - b_7 \alpha_{P_7}^{30} - w_{H}^{30} + w_{H_4}^{30} - U_P + w_{H}^{20} - w_{H_4}^{20} + a_1 U_P + b_1 \alpha_{P_1} \end{Bmatrix} \quad (34)$$

The variation between tolerance surface P_7 and its nominal position related to the datum plane P_1 is defined by the displacement of any point of P_7 compared to the corresponding point of P_1. And this variation only depends on rotation variations, which can be calculated as:

$$\left| 30(-\alpha_{P_7}^{30} + \alpha_{H}^{30} + \alpha_{P_1} - \alpha_{H}^{20}) - 80(\beta_{P_7}^{30} + \gamma_{H}^{30} + \beta_{H_1}^{30}) \right| \le 0.02 \quad (35)$$

• Consider the location tolerance requirement between surface P_1 and P_7. The tolerance requirement being studied is a location of plane P_7 with respect to datum **A** on plane P_1. Unlike the perpendicularity tolerance requirement, the geometrical constraint of location tolerance depends on both rotation and translation variations. In order to respect the location tolerance requirement, the following equations should be satisfied:

$$\left| w_{P_7}^{30} + a_7 \beta_{P_7}^{30} + c_7 \alpha_{P_7}^{30} - v_{H}^{30} + w_{H_1}^{30} + c_1 \alpha_{H}^{20} + a_1 \beta_{H_1}^{30} - c_1 \alpha_{P_1} + 30(-\alpha_{P_7}^{30} + \alpha_{H}^{30} + \alpha_{P_1} - \alpha_{H}^{20}) - 80(\beta_{P_7}^{30} + \gamma_{H}^{30} + \beta_{H_1}^{30}) \right| \le 0.015 \quad (36)$$

6. Conclusions

In the context of tolerance transfer in process planning, our interest focuses onto 3D modeling of manufacturing deviations. The proposed model uses small displacement torsors to represent the geometrical deviations induced by the machining operations and positioning dispersions during the successive machining set-ups, and considers each manufacturing set-up as a single mechanism. Thus, the influence of the process planning on the respect of functional tolerances can be modeled as a chain of small displacement torsors. After having obtained all SDT chains according to set-ups and functional tolerances, we can identify which surfaces and which set-up affect the compliance with functional tolerances. Based on the SDT chains, one can evaluate the effect of these variations on the parts and check whether functional tolerances are respected, or determine the manufacturing tolerances according to the functional tolerances. An example proves the validity of the proposed method.

Under the environment of integrated design and manufacturing, our future research works will focus on: The mathematical modeling of 3D manufacturing variations for integrating product and manufacturing process design; the development of the methods and tools for manufacturing tolerance analysis and synthesis in multi-stage manufacturing processes.

7. References

[1] Qin Y, Lu W, Qi Q, Liu X, Huang M, Scott P and Jiang X 2018 Towards an ontology-supported case-based reasoning approach for computer-aided tolerance specification *Knowl-Based Syst.* **141** pp 129-147

[2] Thimm G, Wang R and Ma Y 2007 Tolerance transfer in sheet metal forming *Int. J. Prod. Res.* **45** pp 3289-3309

[3] Jaballi K, Bellacicco A, Louati J, Riviere A and Haddar M 2011 Rational method for 3D manufacturing tolerancing synthesis based on the TTRS approach R3DMTSyn *Comput. Ind.* **62** pp 541-554

[4] Jiang K, Davidson J K, Liu J and Shah J J 2014 Using tolerance maps to validate machining tolerances for transfer of cylindrical datum in manufacturing process *Int. J. Adv. Manuf. Technol.* **73** pp 465-478

[5] Villeneuve F, Lego O and Landon Y 2001 Tolerancing for manufacturing: a three-dimensional model *Int. J. Prod. Res.* **39** pp 1625-1648

[6] Vignat F and Villeneuve F 2003 3D transfer of tolerances using a SDT approach: application to turning process *J. Comput. Inf. Sci. Eng.* **3** pp 45-53

[7] Villeneuve F and Vignat F 2003 3D synthesis of manufacturing tolerances using a SDT approach *The 8th CIRP International Seminar on Computer Aided Tolerancing*, Charlotte, North Carolina, **41** pp 279-290
[8] Villeneuve F and Vignat F 2005 Manufacturing process simulation for tolerance analysis and synthesis In: Bramley A, Brissaud D et al (eds) *Advances in integrated design and manufacturing in mechanical engineering.* Presented at the 5th International Conference on Integrated Design and Manufacturing in Mechanical Engineering, Bath, England, New York: Springer, pp 189-200
[9] Peng H P and Lu W L 2018 Modeling of Geometric Variations within Three-dimensional Tolerance Zones *Journal of Harbin Institute of Technology (New Series)* **25** pp 41-49
[10] Teissandier D, Couétard Y and Gérard A 1999 A computer aided tolerancing model: proportioned assembly clearance volume *Comput.-Aided. Des.* **31** pp 805-817

Acknowledgments
The authors would like to acknowledge the financial support of the National Natural Science Foundation of the People's Republic of China (No. 51575235).

Dynamic Analysis of Carrier Structure of a Marine Underwater Mechanical Arm

Hu Miao[1], Han Jianggui*, Guo Wenyong*

College of Power Engineering, Naval University of Engineering, Wuhan,430033, China

1608481423@qq.com

Abstract. In this paper, the support structure of a certain type of marine underwater manipulator is taken as the object to carry out solid modeling and mechanical properties analysis. At sea level 5, the maximum stress and displacement of the carrier structure are obtained through static analysis, and the first six frequency modes, peak response frequency and amplitude of the carrier structure are obtained through modal analysis and harmonious response analysis. The results show that the safety can be guaranteed under 5-level sea conditions, and the influence of cyclic loads on the structure is significant. The influence of cyclic loads on the structure should be taken into account when analyzing the structure. The research results have engineering application value and guiding significance for optimum design and safety strength checking of underwater manipulator carrier structure.

1. Introduction

With the rapid development of underwater inspection and maintenance technology of ships in our country, underwater vehicle, as an intelligent device, can replace manual underwater special operation. Its application prospect is very broad. The underwater manipulator provides a feasible way to detect the working conditions of underwater parts of ships, implement underwater emergency repair and replace underwater parts. At present, the technology of underwater working robots in the United States, Japan, Britain, France and other countries is the most developed, most of which are used in underwater maintenance, manned deep submersible and deep-sea working underwater workstations. Therefore, how to realize the transportation, positioning, operation and recovery of the manipulator in the ocean is an urgent problem for all countries under the condition of ensuring the ship's sailing rate and reducing the maintenance cost as far as possible.

In this paper, a carrier structure of underwater manipulator with triangular truss structure as the main body is designed, which can carry and transport the underwater manipulator. SOLIDWORKS is used to build the three-dimensional model of the underwater manipulator carrier structure, and ANSYS is used to carry out static and modal analysis of the structure to evaluate the safety of the underwater manipulator carrier structure. Studying the steady-state response of the structure can help designers overcome resonance, fatigue and other adverse effects caused by forced vibration, and provide technical support for the safe operation of the structure [1].

2. Structure and working principle of marine underwater mechanical arm carrier

The carrier of ship's underwater mechanical arm is mainly composed of triangular truss, working platform, lifting device, suction cup and swinging device, etc., which is mainly used for carrying and transporting underwater mechanical arm, and the end mechanical arm is used to carry many corresponding auxiliary equipment. To realize underwater observation, inspection and simple maintenance of the hull and appendages below the waterline when the ship stays on the water. The overall structure is shown in figure 1.

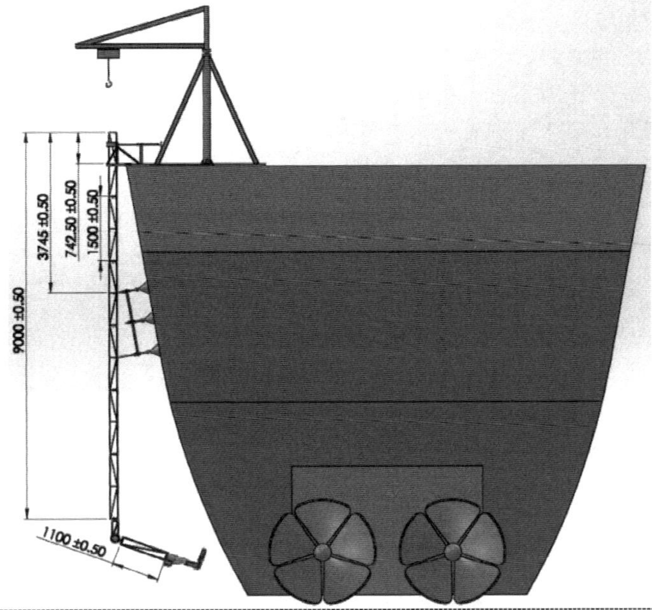

Figure 1. Overall structure.

The structure adopts locking working platform, which consists of main platform and expansion platform. The truss structure is composed of standard section, which adopts triangle structure. Some degrees of freedom of truss structure can be restricted by vacuum suction cup adsorbed on the outside side of ship. The swing device consists of a rotating mechanism and an electric push rod, which can rotate around the bottom end of the truss structure and is used to connect the carrier structure with the mechanical arm. The triangular truss structure is made of 5052 aluminum alloy with yield strength of 195MPa and vertical height of 9m.

3. Environmental load

Under the interaction of random loads such as gravity, wave, sea breeze and ocean current, and drag force and inertia force, the carrier structure of underwater mechanical arm generates dynamic response. In this paper, stokes fifth-order wave theory and Morison wave force are adopted to simulate wave load.

3.1. Stokes wave theory

The waveform of Stokes wave theory is an irrotational wave with periodic fluctuations on its surface. The ratio of effective wave height to wavelength (i.e. wave steepness) H/L is the main factor determining the wave properties [2]. The expression of Stokes fifth-order wave velocity potential is as follows.

$$\varphi = \frac{c}{k}\sum_{n-1}^{c} \lambda_n ch(nk)(z+h)\sin(n\theta) \tag{1}$$

In the formula, "c" is velocity, m/s;"k" is wave number;" λ_n "is known composite function; "z" is height of wave in depth direction, m; "h" is depth, m; " θ " is phase angle, $\theta = kx - \omega t$,where "t" is time (s) and " ω "is wave frequency (rad/s).

3.2. *Drag force around flow*

The drag force on a cylinder of unit length can be calculated by the following formula.

$$f_D = \frac{1}{2} C_D \rho A \upsilon_0^2 \tag{2}$$

In the formula," ρ "is the density of the fluid, g/cm^3; "A" is the projection area of the cylinder perpendicular to the flow direction of the fluid, m^2; "D" is the diameter of the cylinder, m; " C_D "is the drag coefficient, which reflects the viscous effect of the fluid, which is related to Reynolds number and the roughness of the cylinder; ""is the velocity of the fluid, m/s.

3.3. *Inertia force around flow*

The additional mass around the cylinder which changes its original motion will also produce an additional inertial force along the direction of fluid flow. Therefore, the inertial force acting on the cylinder by the accelerated fluid along the direction of flow is as follows.

$$f_I = (M_0 + M_W)\frac{d\upsilon}{dt} = C_M M_0 \frac{d\upsilon}{dt} \tag{3}$$

In the formula, " C_M " is the inertia force coefficient, which reflects the additional mass effect caused by the change of velocity of flow field around the cylinder due to the inertia of fluid and the existence of cylinder [3]." C_m "is the additional mass coefficient, $C_M = 1 + C_m$.

3.4. *Wave force*

In calculating wave forces acting on slender cylinders (D/L $\leq$ 0.2), the semi-theoretical and semi-empirical formula Morison equation [4], which is based on flow theory, is still widely used in engineering design, including two components of horizontal drag force and horizontal inertia force.

$$f = \frac{1}{2} C_D \rho_w D(u - \frac{\partial y}{\partial t}) | (u - \frac{\partial y}{\partial t}) | + C_M \rho_w \frac{\pi D^2}{4} (\frac{\partial u}{\partial t} - \frac{\partial^2 y}{\partial t^2}) \tag{4}$$

In the formula," f " is wave force per unit length, kN/m; " u " is horizontal velocity of water quality point, m/s; " C_D " is resistance factor; " C_M " is inertia force factor; " ρ_W " is sea water density, kg/m^3 ; "D" is pipe diameter, m;" y" is displacement, m; "t" is time, s.

4. Static analysis of carrier structure

4.1. *Establishment of finite element model*

From the above formulas, wave loads and circumfluent loads can be calculated. Assuming that they are in the same direction as the working loads, the results are applied to the truss structure, and the finite element model of the structure is established by using ANSYS. In this paper, only the truss structure is analyzed, without considering the force of suction cup and locking device. The upper end of the truss structure is fixed on the working platform, and all degrees of freedom of the upper end are restrained. The displacement degrees of freedom in X, Y and Z directions and rotational degrees of freedom in Y and Z directions are restrained by the sucker.

4.2. *Nonlinear static analysis of structures*

There are many different working conditions of underwater manipulator in actual working conditions, mainly including positioning, detection, grasping and cutting. This paper mainly considers the self-weight of the structure, the environmental load and the working load of the manipulator when cutting. At this time, the working water depth is 5 meters, and the meaningful wave height is 2.5-4 m under the 5-level sea condition. Take 4 m, the effective period of the wave is 9.09s, and the wave velocity, U= 14.18 m/s, which is the velocity of the water quality point.

The static analysis of the truss structure is carried out under the 5-level sea condition, and the equivalent stress nephogram (Fig 2) and the deformation nephogram (Fig 3) of the whole structure are obtained.From Figs 2 and 3, we can see the Mises stress of the truss structure and the deformation of the truss structure under the 5-level sea conditions.

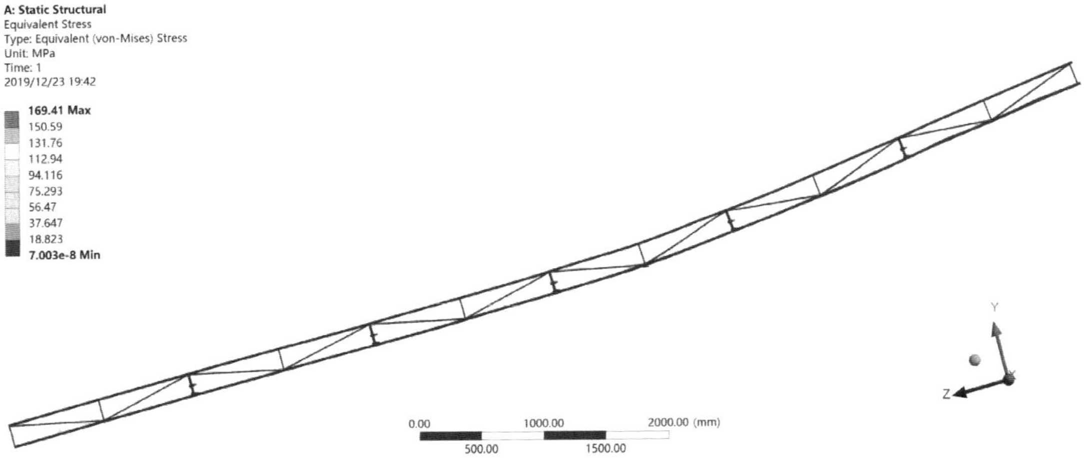

Figure 2. Contour nephogram of equivalent stress.

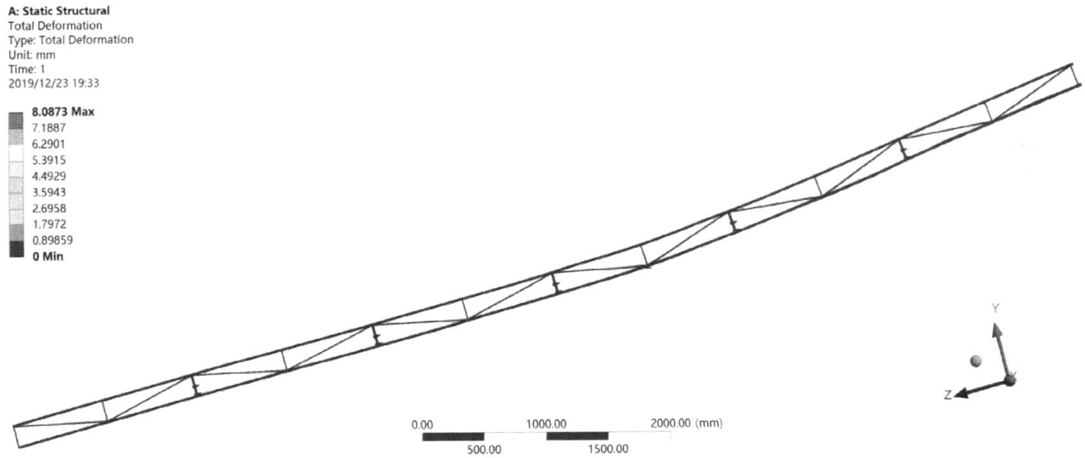

Figure 3. Contour nephogram of deformation.

From Fig 2 and 3,we can see that the maximum stress of the truss structure is 169.41 MPa $<[\sigma]=$ 195 MPa, which is less than the allowable stress. That is to say, the truss structure is safe under the static state under the 5-level sea condition. When the loading direction is perpendicular to the side of the ship, the maximum stress and displacement occur at the bottom of the truss structure connected with the swing device. Therefore, the bottom of the truss structure is the dangerous part of the whole carrier structure. Some improvement measures should be taken to improve the safety and stability of the structure in dangerous places, such as increasing the wall thickness of the dangerous part of the truss structure or replacing the existing materials with higher strength materials.

5. Dynamic analysis of carrier structure

5.1. Structural modal analysis

In this paper, modal analysis is used to solve the natural frequencies and periods of the carrier structure. Compared with the natural frequencies of wave loads, the possibility of resonance of the structure is judged, and the vibration period and natural frequencies of the structure are obtained. The data results are arranged as shown in Table 1 [5].

The natural frequency of the first mode of the structure is 11.059Hz, and the natural vibration period is 0.090s, which is less than the effective period of the structure under the 5-level sea condition of 9.09s. Therefore, the structure can avoid the wave period range under the 5-level sea condition without resonance.

Table 1. The first six natural frequencies and natural vibration periods of truss structures.

Mode	1	2	3	4	5	6
Natural frequency [Hz]	11.059	11.827	40.061	43.350	46.297	47.465
Cycle [s]	0.090	0.085	0.025	0.023	0.022	0.021

5.2. Harmonic response analysis of structures

In this paper, the harmonic response analysis is carried out by means of modal superposition method. By multiplying the factor of mode shape (eigenvector) obtained from modal analysis and calculating the response of the structure, the solution can be gathered near the natural frequency of the structure, and a smoother and more accurate response curve can be generated [6].The analysis results of harmonic response of truss structure is shown in Fig 4 and 5.

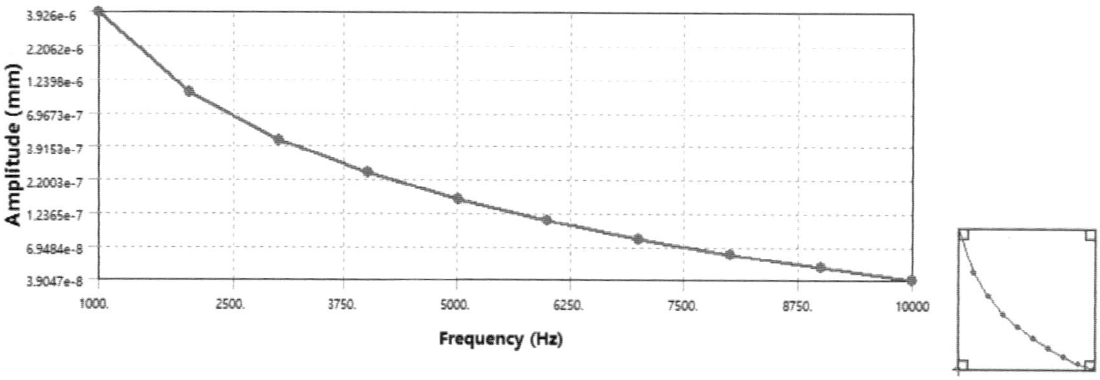

Figure 4. Frequency response diagram of X-direction deformation.

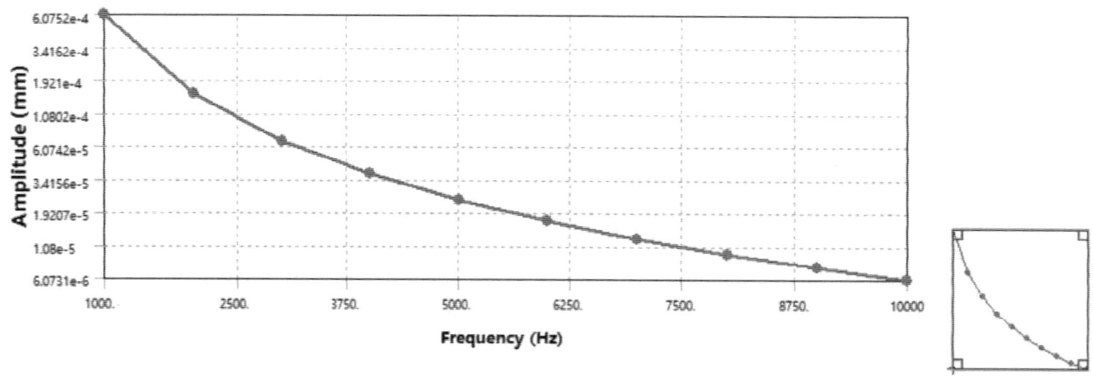

Figure 5. Frequency response diagram of Y-direction deformation.

From Fig 4 and 5,we can see that the vibration displacement changes with the increase of excitation frequency. The peak value appears at 1000 Hz. The maximum vibration displacement in X direction is 3.926×10^{-6} mm (Fig 4) and in Y direction is 6.0752×10^{-4} mm (Fig 5). At the peak excitation frequency, the maximum stress of the structure is 1.423×10^{-4} MPa, which is less than the allowable stress of the structure and meets the design requirements.

6. Conclusion

Taking the underwater manipulator carrier structure as the research object and based on ANSYS software, the strength and harmonic response of the structure under the 5-level sea condition are mainly analyzed. The following conclusions are drawn:

- The maximum stress and displacement of the structure under static loads are obtained. The results show that the structure is relatively safe under 5-level sea conditions.
- When the direction of environmental load is perpendicular to the ship's side, the dangerous part of the truss structure appears at the junction between the bottom of the structure and the swing device, in which certain measures can be taken to ensure the safety of the structure.
- By modal analysis, the natural frequency of the structure is compared with the wave load frequency, and the natural vibration period of the structure is less than the effective period of the wave. It is predicted that the possibility of resonance of the truss structure under the 5-level sea condition is less.
- The harmonic response analysis of truss structure under wave load is studied. The maximum vibration displacement and stress of the structure under excitation frequency are obtained. It is helpful for designers to overcome the harmful conditions such as resonance and fatigue caused by forced vibration.

7. References

[1] Chen Yang, Chen Guoda, Timing Ming, et al. Modal and harmonic response analysis of lightning rod in substation [J]. Electromechanical Engineering, 2017 (2).
[2] Chen Chang, Zhang Shifu, Zhang Qixin, et al. Wave load calculation of offshore floating hose based on improved Morison formula [J]. Journal of Logistics Engineering College, 2015 (2).
[3] Shahai. Resonance Analysis of Thermomixer 5355 Vibration System of Mixer [J]. China Equipment Engineering, 2016 (15): 115-118.
[4] Yu Lixin, Zhuo Yang, Wu Feng. Dynamic simulation sensitivity analysis of offshore wind power jacket foundation [J]. Shanxi Architecture, 2015, 41 (25): 35-36.
[5] Cheng Linfeng, Wang Qian. Modal Analysis and Harmonic Response Analysis of a Ship Control Cabinet [J]. Mine Warfare and Ship Protection, 2016, 24 (1).
[6] Zanyunchao [1], Li Xiangsong [1]. Dynamic analysis of ZS2460 linear vibration sieve [J]. Mechanics and Electronics, 2015 (3): 5-7.

FEM-based Methodology for Crash Severity Estimation in Frontal Crash Scenarios

Gerald Joy Sequeira[1], Akshay Patel[1], Shahabaz Afraj[1], Robert Lugner[1] and Thomas Brandmeier[1]

[1] Research and Test Center CARISSMA, Technische Hochschule Ingolstadt, 85049 Ingolstadt, Germany

E-mail: sequeira@thi.de

Abstract. With the technological development of forward-looking sensors, researchers are exploring their use not only for advanced driver assistance systems but also to gain important pre-crash information. Based on this pre-crash information, if the occupant motion inside the vehicle structure can be predicted for the oncoming crash scenario, then an optimal restraint strategy can be planned before the crash. This paper introduces a two-step FEM simulation based methodology for predicting the occupant severity in head-on crash scenarios. In the first step, we simulate the vehicle level model with different impact positions and relative approach angles. The results of these simulations, linear velocities in the longitudinal and lateral direction and angular velocities (roll, pitch, and yaw) during in-crash phase are the loading conditions for next simulation step (occupant level). This step simulates the motion of the driver in different crash scenarios. In this paper, we investigate the head, neck, and chest injury risks from vehicle-to-vehicle crash both traveling at 50 kilometers per hour. Prediction of the head injury criterion, identifying the cases where additional deployment of side-airbags and discussion of injury criteria with contour plots are the main outcome of this paper.

1. Introduction

Vehicle manufacturers are in continuous search of methods and techniques to reduce fatalities and thus move a step closer to realize Vision-Zero [1]. Forward-looking sensors (radar, lidar, camera, etc.) and their corresponding technology are already in-use for driver-assistance, driver-warning and collision avoidance systems. Pre-crash information such as opponent type, velocity, relative approach angle and position of the first contact define a particular crash configuration, which can be used to activate the restraint systems earlier than present traditional passive safety systems [2]. It might be required to activate restraint systems before the first contact to save the occupants. The decision to activate a restraint system or combination of restraint systems and their activation time is very complex and requires prediction of crash severity based on occupant kinematics during a particular crash configuration.

Crash severity prediction is a vital part of the above described predictive vehicle safety system. Conventional passive safety system also predicts the crash severity based on the data measured by different safety sensors (acceleration-based, pressure-based, structure-borne-sound based). Based on this prediction, the corresponding safety action is deployed. This technology has a limit on activation time of airbags depending on the response time of the passive safety sensors (about 15 milliseconds). With the information from the forward-looking sensors and advanced crash severity prediction, there is a possibility to activate the restraint systems earlier. A number of studies used real-crash database such as Crash Analysis and Reporting System (CARS), National Automotive Sampling System (NASS) and General Estimates System (GES) for training different machine learning algorithms with the aim of predicting crash severity [3, 4]. Some studies used Artificial Neural Networks to identify and predict influential factors contributing to crash severity [5, 6]. The severity of an occupant injury in a particular crash is a combination of the physical behavior of crash management structure, occupant kinematics and the effect of restraint systems. Previous studies have considered vehicle level parameters for either training the model or classifying and predicting the severity. These studies lack the physical behavior of the occupant kinematics and restraint systems. This paper describes an approach to investigate occupant level injury scales (as per Euro-NCAP) under different crash configurations using a two-step FEM simulation. The knowledge gained from the analysis of the occupant injuries under different crash scenarios can be used to decide an appropriate pre-crash restraint strategy (restraint strategy before the crash). This knowledge would also help in investigating the limitations of standard restraint systems and highlight the crash configurations which might need advanced systems such as smart airbags to save the occupants. According to the Insurance Institute of Highway Safety (IIHS) passenger vehicle occupants accounted for 64 percent of vehicle crash deaths and about 17,662 (74.5 percent) passenger vehicle drivers died in the year 2017 [7]. Moreover, the statistics based on the crash-type show that frontal crash scenario accounts for the highest percentage of passenger vehicle occupant deaths (about 56 percent) followed by the side crash scenario at 24 percent [7]. Hence, this research work is primarily focused on frontal crash. Even though this paper discusses the results from investigations of different frontal crash scenarios, this methodology can also be applied for other crash scenarios like side and rear-end crashes.

2. Methodology

A two-step FEM simulation approach as shown in Fig. 1 is used for acquiring the occupant responses in different crash scenarios. In a first step, different crash scenarios based on different pre-crash information are simulated at vehicle level to extract the in-crash vehicle responses. These responses include high frequency components which are filtered with the standard low-pass SAE-J211 filter with 100 Hz cut-off frequency. The filtered data is then fed to a second level FEM-simulation (occupant level simulation) to extract the occupant responses. Based on these occupant responses the injury criteria for different body parts are calculated. In the end, these injury criteria are analyzed for different crash scenarios and their relationship is studied.

For vehicle level simulation, the vehicle model of Toyota Yaris made available by Collision Safety and Analysis (CCSA) of George Mason University was used. The severity for head-on frontal crash is greater as compared to in-line collision. Moreover, most of the frontal collisions occur when driving on city roads, where the speed is limited to 50 kilometers per hour (e.g. Germany). Hence for vehicle level simulation, head-on collision crash scenarios were selected with both vehicles traveling at 50 kilometers per hour. Considering different loading conditions on CMS as discussed in section II-A, 13 different impact positions were selected (6 on driver side, 6 on passenger side and a central impact position). For each position, 11 relative angular cases (5 positive angles, 5 negative angles and a 0 degree angle) with a variation of 15 degrees between each case were chosen for simulation. Considering front curvature of the vehicles, collision at extreme impact position with angular cases is geometrically not possible and hence were not simulated. Based on these conditions, a total of 131 different scenarios were simulated. The results of these simulations, mainly the motion such as linear and angular velocities of vehicle during the crash is then given to the occupant level simulation. Since the driver is more prone to injuries, this study focuses on driver injuries in the occupant-level simulation. The occupant-level simulation includes restraint systems such as driver-airbag and seat-belt with pre-tensioner

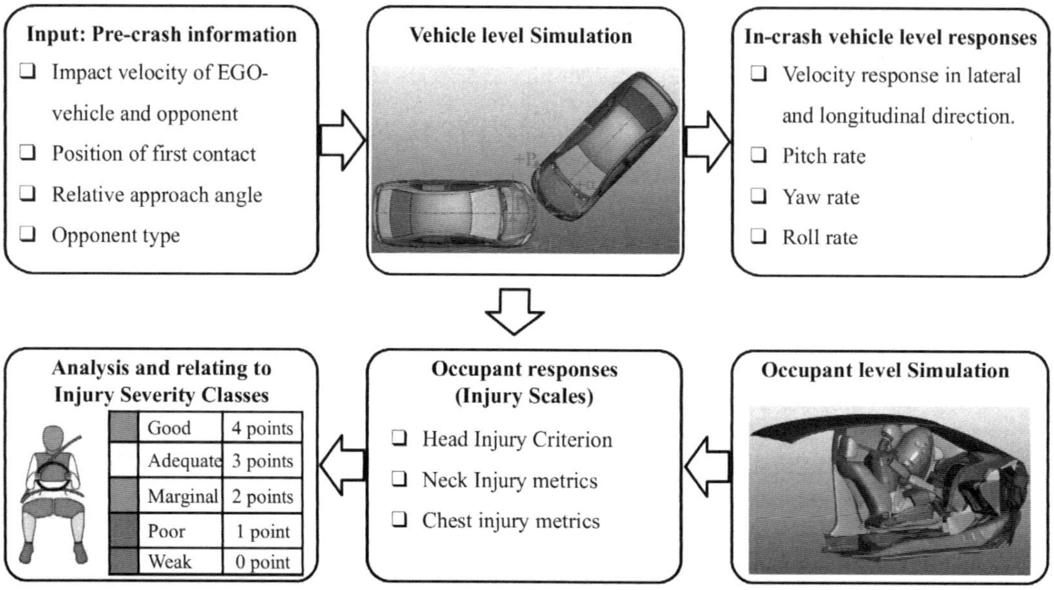

Figure 1. Methodology for data generation and analysis (vehicle level simulation blocks displays the sign conventions used for impact position and relative approach angle)

and retractor. In most of the European vehicles, frontal airbags are installed voluntarily by the manufacturers. We have chosen these restraint systems complying to the least possible configuration of restraint systems available for purchase in Europe. In our study, we activate seat belt retractor at 1 millisecond and is set to a maximum force of 3.25 kilo Newtons, while the front airbag is activated after 15 milliseconds and fully inflated at 45 milliseconds after T0. The restraint system activation parameters were the same for all the occupant-level simulations in order to compare the severity on the occupant body parts.

3. Results and discussion

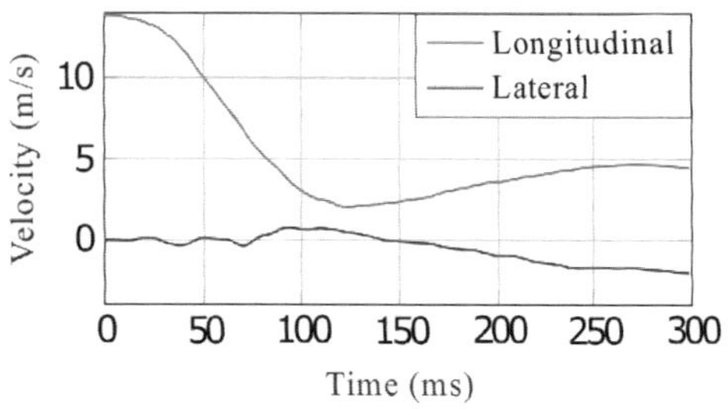

Figure 2. Longitudinal and lateral velocity from vehicle level simulation.

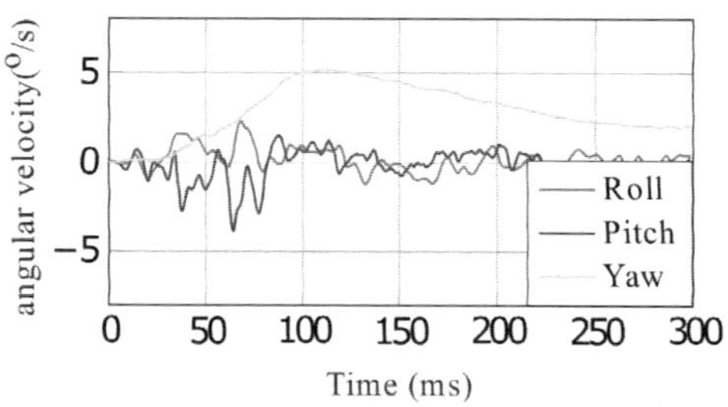

Figure 3. Angular velocity (roll, pitch and yaw rates) from vehicle level simulation.

The results from vehicle level simulations vary with the position of impact and relative approach angle. The simulations with both impact position and relative approach angle near or equal to zero show very little increase in lateral velocity, yaw rate and roll rate. As the impact position or angle changes, the vehicle moves laterally along with angular motions. Fig. 2 and

Table 1. Parameters of curve fit for different impact positions

Position			Parameters of Gaussian fit			Goodness of fit
Number	Distance from center (mm)	Normalized position	a	b	c	R^2
P3	364	0.43	727.20	11.38	23.83	0.97
P2	245	0.29	362.70	45.20	66.21	0.98
P1	122	0.14	341.20	33.44	67.69	0.97
P0	0	0.00	335.90	15.75	49.35	0.98
-P1	-122	-0.14	324.40	12.32	50.83	0.95
-P2	-245	-0.29	326.00	6.33	49.66	0.93
-P3	-364	-0.43	224.50	5.14	60.16	0.94
-P4	-485	-0.57	196.30	-9.95	55.81	0.89

Table 2. Coefficients for equation of Gaussian curve parameters

Coefficient number	Parameter of Gaussian equation		
i	a	b	c
1	5605	-1422	-458.4
2	8350	-1665	-1383
3	2798	-158.8	-550.3
4	-788	256.1	192.6
5	-139.5	86	75.2
6	344.3	17	52.9

Fig. 3 show the vehicle responses during crash from one of the crash scenario simulated (impact position at 485 mm from the center at an angle of -45 degree).

The upper-body (head, chest and neck) of a human is more sensitive and prone to fatal injuries than the lower-body (femur, tibia, knee and feet). Therefore, we are focusing our study on the behavior of upper-body injuries with different crash scenarios. The following section discusses the results from occupant level simulation.

3.1. Head

The first criterion considered for head injury is $A-3ms$. As per the EuroNCAP norms, a hard contact of head is defined by a value higher than or equal to 80 g. Fig. 4 shows a contour plot of this criterion with different positions and angles. It can be observed that a hard contact occurs at extreme positions (normalized position above 0.57) and angles as shown by red region in the contour plot. There are two cases for a hard contact. The first case consists of the crash scenarios causing the driver to move diagonally towards the A-pillar and side structures. In this case, fatal injuries are highly probable due to direct contact of head with the structures. The other case includes the driver diagonal movement towards the airbag. The driver's head in this case contacts with airbag at an offset position (out-of-position) and does not have the required stiffness to stop further movement of the head. Therefore after some time, the head comes in contact with steering wheel through airbag cloth causing a hard impact.

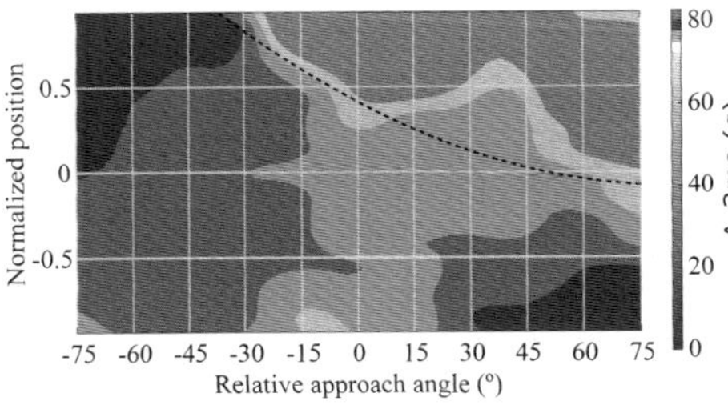

Figure 4. $A-3ms$ contour plot (color map of plot is as per EuroNCAP injury rating scheme [8].

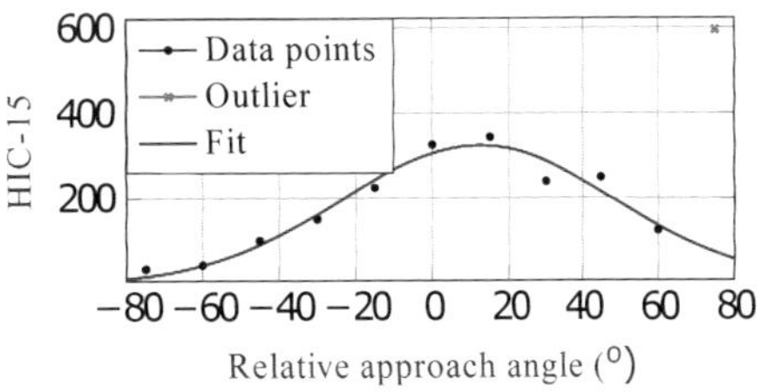

Figure 5. Fitted curve for HIC-15 data points (normalized position=-0.144).

$HIC-15$ is the second criterion considered. The value of this criterion for a particular impact position is high at angles near zero and reduces with increase in the angle (both positive and negative increment) and can be represented by a Gaussian equation

$$HIC = a \cdot e^{-\left(\frac{\alpha-b}{c}\right)^2} \tag{1}$$

where a, b and c are the real arbitrary parameters of the Gaussian curve and α is the relative approach angle in degrees. Hence, to analyze the behavior of this criterion, a number of Gaussian curves with each curve corresponding to a particular position was fitted. Fig. 5 shows the fitted curve for a particular impact position. The crash scenarios with extreme impact positions include large number of data points which fall in the first case of hard contact. Hence, the data points for these impact positions were not considered for curve fitting analysis. For the impact positions, where very few data points represent hard contact were treated as outliers and hence neglected during the curve-fitting process. Table 1 displays the results from the Gaussian curve fitting process for different impact positions. The behavior of these parameters obtained was approximated by polynomial equations as given below

$$a = d_1 \cdot P^5 + d_2 \cdot P^4 + d_3 \cdot P^3 + d_4 \cdot P^2 + d_5 \cdot P + d_6 \tag{2}$$

$$b = e_1 \cdot P^5 + e_2 \cdot P^4 + e_3 \cdot P^3 + e_4 \cdot P^2 + e_5 \cdot P + e_6 \tag{3}$$

$$c = f_1 \cdot P^5 + f_2 \cdot P^4 + f_3 \cdot P^3 + f_4 \cdot P^2 + f_5 \cdot P + f_6 \tag{4}$$

where d_i, e_i and f_i are the coefficients of the polynomial equation and P is the normalized position of impact. Substituting the above polynomial equations in (4), we get

$$HIC = (d_1 \cdot P^5 + d_2 \cdot P^4 + d_3 \cdot P^3 + d_4 \cdot P^2 + d_5 \cdot P + d_6) \cdot e^{-\left(\frac{\alpha - e_1 \cdot P^5 + e_2 \cdot P^4 + e_3 \cdot P^3 + e_4 \cdot P^2 + e_5 \cdot P + e_6}{f_1 \cdot P^5 + f_2 \cdot P^4 + f_3 \cdot P^3 + f_4 \cdot P^2 + f_5 \cdot P + f_6}\right)^2} \tag{5}$$

This equation represents the surface of $HIC - 15$ for different positions and angle, which is valid for limits $P >$ -0.57 to $P <$ 0.43. Table 2 gives the coefficients of the polynomial equations obtained by curve fitting. $HIC - 15$ for positions beyond the limits of the above equation is very high due to the direct contact of the occupant with side-structures or improper position of contact with airbag (out-of-position). This highlights the need of other airbags such as curtain airbag or torso airbag to reduce the probability of fatal injuries to the occupant.

3.2. Neck

NCAP-protocol specifies time-based cumulative limits for neck forces. In our study for all the crash scenarios, the neck forces were in the safe region (green) up to time corresponding to the end of the cumulative limits (45 milliseconds for shear force and 60 milliseconds for tension) [9, 8]. Beyond this time, the neck forces can only be classified as good (green region: when the force is below 1.1 kilo-newtons) or poor (red region: when the force is equal to or above 1.1 kilo Newtons). Fig. 6 shows the contour plot for neck shear forces. It can be observed that as the normalized overlap position increases along with the relative approach angle, the neck force changes from good region to fatal (poor) region. The primary cause for this transition is the direction of resultant velocities acting on occupant body. Due to smaller overlaps and higher approach angles, the lateral velocity, yaw, pitch and roll components increase in magnitude and cause the occupant dummy to move in a direction that reduces the effective interception with airbags. In some cases, this leads to a hard contact with internal car structure and results in high neck forces. However, in some cases it causes the occupant body to change the direction of motion abruptly and due to inertial effect between head and the body, the forces on neck increase. The neck tension force and extension moment show similar behavior as that of the neck shear force.

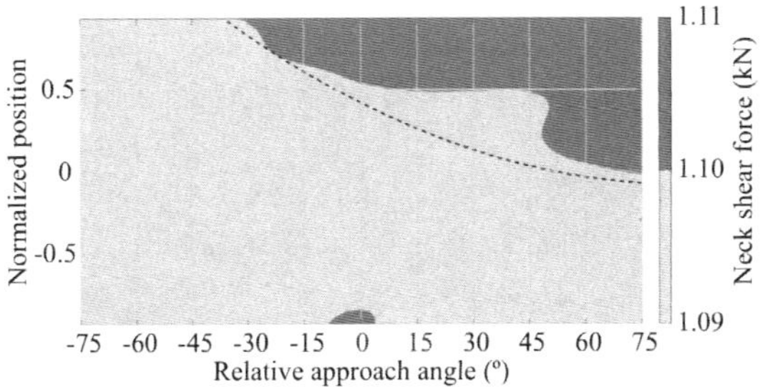

Figure 6. Neck shear force contour plot.

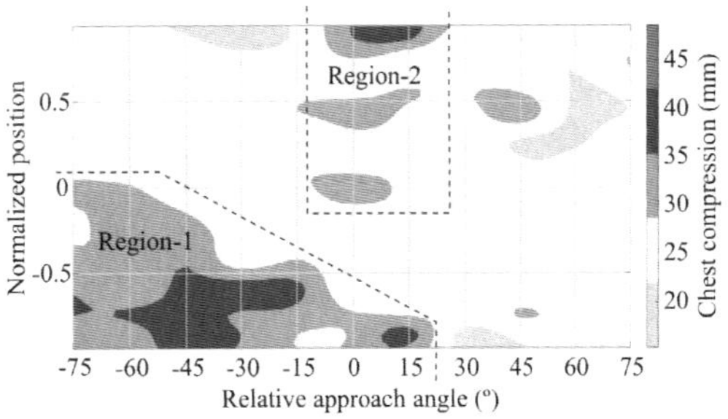

Figure 7. Chest compression contour plot.

Chest compression during a crash mainly depends on the effective impact on the vehicle and the mechanics of the restraint systems relative to motion of the occupant. This behavior can be observed in the contour plot (Fig. 7). There are two cases which cause an incremental transition in the injury values based on chest compression.

Case-1: This case consists of crash scenarios with extreme negative positions and negative approach angles (Region-1 in Fig. 7). The increase in chest compression is caused due to the relative dynamics of vehicle and occupant during the crash. The lateral velocity and rotational motion induced due to the combination of positions and angles of vehicles during crash, cause the occupant to move in directions that would not effectively intercept the inflated airbag. Hence the damping efficiency is reduced, and the occupant moves a larger distance before beginning the rebound phase. This causes the seat belt to provide higher reaction forces onto the chest of the occupant, which increases the chest compression.

Case-2: This case consists of crash scenarios with positive positions and relative approach angle near zero degree (Region-2 in Fig. 7). For these crash scenarios, we observe that the effective energy transferred during the crash is maximum. Hence the occupant velocity would be higher. The restraint systems have higher forces to dampen, to decelerate the occupant. As a result, the overall deflection of the chest increases.

4. Conclusion

Following points can be concluded based on the analysis of the results carried out in section IV.

- In crash scenarios with extreme positive positions and positive relative approach angle, the occupant's head and neck have a higher risk of fatal injuries. For these crash scenarios, activating side airbags (if it's installed in the vehicle) would reduce the occupant injuries. A safest possible boundary line can be defined based on the contour plot for both head and neck injury criterion ($A - 3ms$ and neck forces). An arc representing the boundary of the fatal region can be used to predict the possibility of a fatal crash based on the angle and position of crash (shown by a dashed arc in Fig. 4 and Fig. 6). Based on the simulations conducted in the scope of this research, this boundary arc can be represented by following equation
$$(\alpha - 85.35)^2 + (P - 7283.02)^2 = 7283.1^2 \qquad (6)$$
This arc would help to distinguish between the crash scenarios which require activation of frontal airbags only and those which require activation of side curtain airbag and torso airbag additionally.
- For the crash scenarios with proper contact with airbag, a methodology to predict the HIC-15 values based on the position of impact and relative approach angle is formulated. This

methodology would help in predicting the head injuries in a particular crash scenario before the crash.

- It can be observed that the higher risk region is developing in the extreme negative positions and negative angles for chest compression. A prior understanding of the transitional behaviour can be used to trigger the torso airbags (if equipped) to restrict the motion of the occupant and hence mitigate the possibilities of higher chest deflection.

5. Future work

In this paper, we have analyzed the different frontal head-on crash scenarios with both the vehicles traveling at 50 kilometers per hour. The study can be extended to different velocities and opponents to get information about the occupant injuries by these parameters. This information can be used to build an occupant injury prediction model comprising of different models for each criterion. This prediction can help in deciding which combination of the restraint systems to be activated and at what times. These steps are planned as future research work.

References

[1] European Union. *Mobility and transport - Road Safety.* URL: https://ec.europa.eu/transport/road_safety/home_en.

[2] G. J. Sequeira et al. "A validation sensor based on carbon-fiber-reinforced plastic for early activation of automotive occupant restraint systems". In: *Journal of Sensors and Sensor Systems* 8.1 (2019), pp. 19–35. DOI: 10.5194/jsss-8-19-2019. URL: https://www.j-sens-sens-syst.net/8/19/2019/.

[3] Miao M. Chong, Ajith Abraham, and Marcin Paprzycki. "Traffic Accident Analysis Using Machine Learning Paradigms". In: *Informatica (Slovenia)* 29 (2005), pp. 89–98.

[4] Abhishek Das, Mohamed Abdel-Aty, and Anurag Pande. "Using conditional inference forests to identify the factors affecting crash severity on arterial corridors". In: *Journal of safety research* 40.4 (2009), pp. 317–327. DOI: 10.1016/j.jsr.2009.05.003.

[5] F. Rezaie Moghaddam, Sh. Afandizadeh, and M. Ziyadi. "Prediction of accident severity using artificial neural networks". In: *International Journal of Civil Engineering* 9.1 (2011). URL: http://ijce.iust.ac.ir/article-1-544-en.html.

[6] Sharaf Alkheder, Madhar Taamneh, and Salah Taamneh. "Severity Prediction of Traffic Accident Using an Artificial Neural Network". In: *Journal of Forecasting* 36.1 (2017), pp. 100–108. ISSN: 02776693. DOI: 10.1002/for.2425.

[7] Insurance Institute for Highway Safety. *Fatality facts.* URL: https://www.iihs.org/iihs/topics/t/general-statistics/fatalityfacts/passenger-vehicles.

[8] Euro NCAP. *Assesment protocol - adult occupant protection.* February 2019. URL: https://cdn.euroncap.com/media/43372/euro-ncap-assessment-protocol-aop-v901.pdf.

[9] Global NCAP. *Assesment protocol - adult occupant protection.* August 2017. URL: http://www.globalncap.org/wp-content/uploads/2018/03/assessment-protocol-Adult.pdf.

Acknowledgment

The authors appreciate and thank Center for Collision Safety and Analysis (CCSA), George Mason University, Virgina, USA for providing the validated vehicle model for research use. This work is supported under the funding program Forschung an Fachhochschulen of the German Federal Ministry of Education and Research (BMBF) under contract number 13FH7I03IA (SAFIR IP3). We thank the German Federal Ministry of Education and Research (BMBF) for funding the project.

Product Life Cycle Analysis of Automotive Products on Multi-Purpose Vehicle's (MPV) Segment

A Suzianti, D R Putra and C P Sibuea

Universitas Indonesia, Kampus Baru UI Depok, Jawa Barat, 16241, Indonesia

suzianti@eng.ui.ac.id

Abstract. Since 1990, the ever-increasing sales trend of automobiles in Indonesia has made the competition atmosphere much fiercer, particularly for Multi-Purpose Vehicle, the car segment with the highest sales in Indonesia's automobile market. Therefore, a product life cycle management of automobiles is demanded to secure current position in market or further, to lead the market. This research uses MPV Model A and MPV Model B as a study case for Multi-Purpose Vehicle. Methods deployed are the computation of product line extension, the computation of product line innovation, t-test, Person Index Test, and Importance-Performance Analysis. Results are used to formulate product life cycle management strategies for automobiles of Multi-Purpose Vehicle aimed secure current position in market or further, to lead the market.

1. Introduction

Car sales growth into one of the parameters that indicate economic growth and growth in the industrial sector due respect to purchasing power. Car sales growth in Indonesia is one of the biggest market in ASEAN region, successfully sold 940,000 units over Thailand only 698,000 units in 2015. There are 4 cycles of growth in car sales in Indonesia.

In the first cycle, from 1993 to 1998. The highest achievement in this cycle in 1997 with the total sales of 389 thousand units. The decline that occurred due to the global economic crisis that affected Indonesia and affect the political situation as well. In the second cycle, between 2000 and 2006. The decline that occurred in 2006 was also caused by the global economic crisis. The same thing happened in the third cycle from 2006 to 2009, and 2010 to 2015. So, it is clear that the decline in car sales in Indonesia is largely determined by economic growth in Indonesia, and also global economic.

The purpose of this study is to analyse the product life cycle of car with the MPV segment represented by MPV Model A and MPV Model B by using the method of calculation of Product Line Innovation, and Product Line Extension, and the Importance-Performance Analysis to analyse the facelift and the development of variants on both models. That may eventually provide output for both products on their development strategy in the automotive product lifecycle management on MPV segment.

2. Study of Literature

2.1. Product Life Cycle Management

Levitt (1965) first introduced the concept of the Product Life Cycle (PLC) as a predictive instrument to do forecasting marketing needs and assist in planning their long-term product strategy. PLC is used for the company's strategy, product development, finance, and production, and is considered to be an influential concept.

PLC curve is described as a bell-shaped curve that describes the journey of products since its product introduction into the market until its withdrawal from the market. Model hypotheses showed that sales follow the sequence of steps, beginning with product introduction and continued with growth and maturity, and decline.

The term 'life cycle' generally indicates a whole series of stages, which could be recognized as an independent stage for continued or followed or implemented by a product since its 'birth' to expire. (Terzi et al., 2010). By adopting a model that is easier to use, the product life cycle can be defined by three main phases (Kiritsis, 2003), namely: Beginning of Life (BOL), Middle of Life (MOL), and End of Life (EOL).

(Rogers, 1962) Stating that the number of sales of a product from beginning until the end, can be the reference of trend that existed. Other significant factor also the instalment of specific feature in the existing model. The effectiveness of product lifecycle management policy is very important in the automotive industry related to consumer saturation, and assisting planning strategy of product changes from existing model to new model products. (Volpato, Stocchetti, 2008). Therefore, product life cycle management is crucial for the success of automotive products to compete in a Indonesia's market.

2.2. Product Line Extension (PLE)

Product line extension is one of the indicators used to analyze product life cycle management (Volpato; Stocchetti, 2008). Product line extension is when a company creates a new product in the same product line of existing brands. It can be done with the product strategy to distinguish shapes, different colours, sizes, and have different ingredients or served in various flavours (Kotler, 1997).

2.3. Product Line Innovation

Product line innovation is also an indicator in the analysis of the product lifecycle management. This indicator based on the renewal and additional variants of a product. The results of the calculation of product line innovation will not represent how innovative a new model compared to the old model, but rather shows how often a brand renew or innovate its products (Volpato; Stocchetti, 2008).

2.4. Importance-Performance Analysis

Importance-Performance Analysis (IPA) was first introduced by Martilla and James (1977) as a framework to understand consumer satisfaction as a function of expectations related to the important attributes and research that related to performance. Although each provide the necessary information, the full potential of the information above realized when two concepts were combined.

The first methodology focuses on identifying performance gaps, which are usually obtained by calculating the difference between performance and importance. The second approach to the IPA involves mapping the average yield of the two points on a two-dimensional graph (called "Action Grid" by Crompton and Duray) to make a four-quadrant matrix that identifies areas that need improvement and areas with the effective performance

Quadrant I (High Importance / Low Performance) is named as "Concentrate here". Elements that fall in this quadrant represents a core challenge that requires direct corrective action and should be given the highest priority. Quadrant II (High Importance/ High Performance) is named as "Keep up the good work" contain elements that are a strength for an organization and should be maintained. If the element is located in quadrant III (Low Importance / Low Performance) does not represent a threat to an organization, the dismissal of the resource / business can be done. The quadrant is named as "Low priority". Quadrant IV (Low Importance / High Performance), which is named as "Possible overkill" contain elements which constitute a significant force for an organization and shows the area where resources can be diverted to another place.

3. Methodology

This study is an analysis that was developed based on the product life cycle management. At the stage of data collection is done by taking the data from the company that made the object of research, then at a later stage by doing the calculation of PLE and PLI, and then the next stage is to do a significance test and correlation tests, and at the final stage of the analysis done by using importance-performance analysis.

In the first stage, or the data collection, data is extracted from historical data that is owned by the company regarding the research object which is MPV Model A and MPV Model B. In addition, as the supporting data that is a version of the product, prices, and specifications, taken from books, the internet, and journal or similar research.

In the second stage, after collect the historical data from 2004 to 2015, then calculate the product line extensions and product line innovation. The results of this calculation can describe how big the development or differentiation of products are made in the MPV Model A and MPV Model B throughout their life cycle. And also based on the calculation of PLI will describe how often a product done their facelift or innovation.

In the third stage, after calculation of product line extension and product line innovation from both of the object of study, then do the significance test and correlation test between PLE by the number of sales and PLI by the number of sales to determine which variables that have more influence and relationship to the amount of sales that has been recorded.

The fourth stage or the final stage, after receiving the results of significance test and correlation test, then analysed using the importance-performance analysis to make the right strategy in developing products on the MPV segment. This analysis begins with formulating the action items based on the variables that are taken in the calculation of PLE and PLI, and also the action items is very possible for the company to do to the products that made the object of this study.

4. Data and results

4.1. Stage 1: Data Collection

From the table shows the number of versions contained in the MPV Model A. The meaning of this number of versions is the number of variants offered by XYZ Company to consumers. Examples such as MPV Model A that offers a variant E, G, and V in which each of these variants has significant difference. Starting from the installed features to the price offered. Variant V on MPV Model A offered several sub-variants again with different engine and transmission. For example, MPV Model A V 2.0 A / T. The meaning of such variants is a unit of MPV Model A with a V variant/version with 2000 cc of petrol engine and an automatic transmission. While that fact XYZ offers for variant V itself is consisted of two engines and two transmissions. Namely 2000 cc for petrol engine and 2500 cc for diesel, as well as automatic transmission and manual transmission which with all versions have been offered at a price that varies according to the engine option and transmission option chosen by the consumer.

Also, at this stage is collecting the same data for MPV Model B, as well as sales data for MPV Model A and MPV Model B.

Table 1. Number of Version of MPV Model A.

Years	2004	2005	2006	2007	2008	2009	2010	2011	2012	2013	2014	2015	2016
Total Variant	10	11	10	10	20	16	16	17	17	18	19	17	13

4.2. Stage 2: Calculations of PLE, PLI, and Price Range

MPV Model A. From the figure 1 showing the results of calculation of product line extensions, product line innovation, and price range. It can be concluded that the visible link between PLE with PLI because of fluctuations in the PLE looks similar to the fluctuations that occur in PLI. For a chart of price ranges, seen stagnant from year to year, or the fluctuations are narrow, it shows that the increasing number of variants does not mean the growing magnitude of the price range between the entry-level variant to the flagship variant.

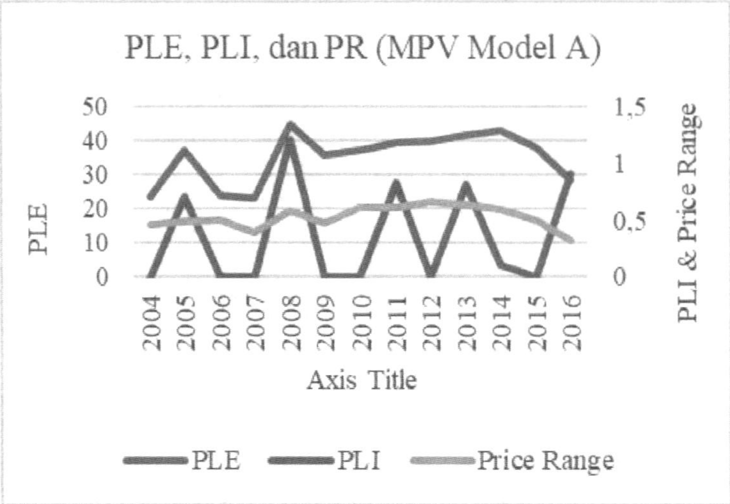

Figure 1. PLE, PLI, and PR (MPV Model A)

MPV Model B. From the figure 2 showing the results of calculation of product line extensions, product line innovation, and price range. It can be concluded that the visible link between PLE with PLI because of fluctuations in the PLE looks similar to the fluctuations that occur in PLI. For a chart of price ranges, seen stagnant from year to year, or the fluctuations are narrow, it shows a same thing with MPV Model A that the increasing number of variants does not mean the growing magnitude of the price range between the entry-level variant to the flagship variant.

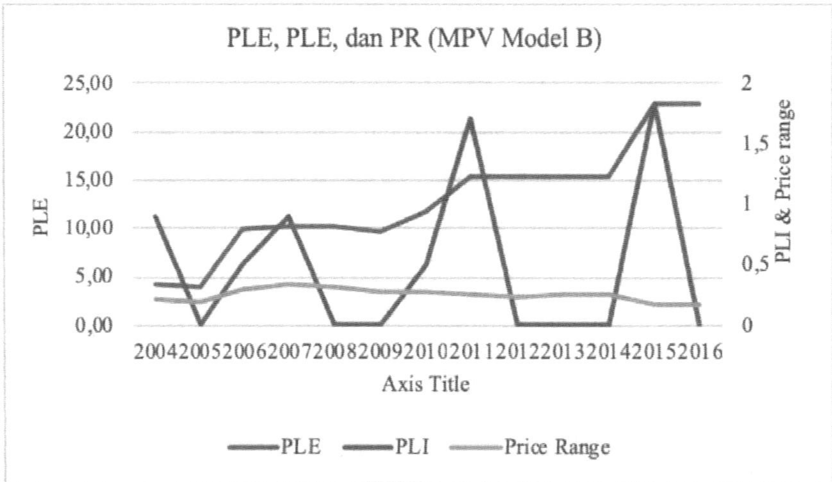

Figure 2. PLE, PLI, and PR (MPV Model B)

4.3. Stage 3: Significancy Test and Correlation Test.

Table 2 shows the results of tests of significance (t-test) and correlation test (Pearson Index) between the calculation of PLE, PLI to the total sales of MPV Model A. From the results generated that shows the p-value of 0.011, and the numbers Pearson correlation of 0.675 may concluded that the product line extension or can be explained as well as product differentiation, is more influence on the amount of sales when compared with the product line innovation. Product line extension also has a correlation that is stronger against the number of sales. Therefore, the conclusion that can be drawn from the results of tests of significance and correlation test, is the development of product differentiation will be more important than innovation or facelift that implemented on automotive products on Multi-Purpose Vehicle (MPV) segment (this calculation is for MPV Model A).

Table 2. Results of t-test and correlation test on MPV Model A

	P-Value	Pearson Correlation
PLE -> Sales	1.143	0.285
PLI -> Sales	1.143	0.285

Table 3 shows the results of tests of significance (t-test) and correlation test (Pearson Index) between the calculation of PLE, PLI to the total sales of MPV Model B. From the results generated that shows the p-value of 0.339, and the numbers Pearson correlation of 0.288 may concluded that the product line extension or can be explained as well as product differentiation, is more influence on the amount of sales when compared with the product line innovation. Product line extension also has a correlation that is stronger against the number of sales. Therefore, the conclusion that can be drawn from the results of tests of significance and correlation test, is the development of product differentiation will be more important than innovation or facelift that implemented on automotive products on Multi-Purpose Vehicle (MPV) segment (this calculation is for MPV Model B).

Table 3. Results of t-test and correlation test on MPV Model B

	P-Value	Pearson Correlation
PLE -> Sales	0.339	0.288
PLI -> Sales	0.912	0.034

4.4. Stage 4: Formulation of Action Items and Importance-Performance Analysis

Based on the results of significancy test and correlation test, the tentative conclusion that can be drawn is that the product line extension is much more influence and has more correlation to the number of sales when compared to the product line innovation.

By looking at the factors involved in the calculation of product line extension which are price range, variants and engine options, there are three things that will be the action items from the calculation of product line extension, as follows:

- Growing or expanding the number of versions in a model that is offered in every one or two years for maximum (Attribute 1)
- Adding and developing a hybrid engine option for the MPV segment (in this case is MPV Model A or MPV Model B) (Attribute 2)
- Developing an engine with green technology or environmentally friendly that will be carried on MPV Model A and MPV Model B in the future. (Attribute 3)

Furthermore, by looking at the factors that are involved in the calculation of product line innovation that shows how much innovation or facelift is implemented and how often innovation or facelift is done, then there are two things that will be the action items from the calculation of product line innovation, as follows:

- Doing an innovation for each model every one year or two years for maximum (Attribute 4)
- Doing a *facelift* for each model every one year or two years for maximum (Attribute 5)

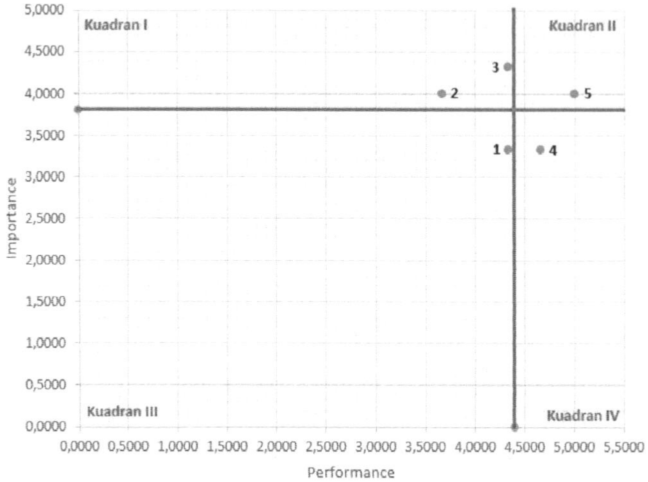

Figure 3. Quadrant Results of Importance-Performance Analysis

In quadrant one, there are two action item that is considered to have a high level of importance with low performance. These attributes are attribute 2 and 3.

Developing hybrid engine options and engine with green technology for MPV Model A or MPV Model B has a high importance to the company and the environment due to fuel efficiency goals for every liter of gasoline used in a vehicle. Values of importance to attribute 2 is equal to 4 and attribute 3 is 4.33. So, it shows the importance of the development of hybrid engines and engine options with a green technology.

If we looked from the low performance, it can be concluded that it happened due to lack of consumer demand that is strong enough regarding the development of their options of hybrid engines or engine with a green technology for the MPV Model A and MPV Model B, which would certainly affect the price of products yet the products are still considered to be efficient or have an economical fuel by Indonesian consumers. Value of performance for attribute 2 is 3.67 and attribute 3 is about 4.33

In Quadrant two, there is one action item that is considered to have a high performance, so it is worth to maintain its performance. Attributes inside this quadrant is attribute 5. The facelift always maintained with a period of approximately one and a maximum of two years in order to meet the needs of consumers and strong enough to compete with the competitors. Importance value of attribute 5 is 4, and performance value of attribute 5 is 5.

In quadrant three, there is one action item that entered in a low importance and low performance: Growing or expanding the number of versions in a model that is offered in every one or two years for maximum (Attribute 1)

Attribute that is inside quadrant four is considered as a not important but has a good performance. In this quadrant, there is one action items which is Attribute 4. It is considered to have a lower importance than other attributes due to the innovation that is certainly needs research and development costs which are not cheap, so it does not make their importance becomes high. But with their innovations will affect the number of sales so it can be concluded that it has a high-performance value. Performance value for the attribute 4 is 4.67, and for the importance value is 3,33.

5. Conclusion

This study aims to recommend the product life cycle strategy for XYZ company. This study gives feedback for product life cycle management, not financial feasibility study and marketing model. From data processing and data analysis results, it can be concluded that the importance of product lifecycle management evidenced by the relationships and correlations between product line extension and product line innovation with sales amount of MPV Model A and MPV Model B.

From the action items that have been formulated, suggestions for priority implementation strategies of XYZ company in their product life cycle management for multi-purpose vehicle segment, Concentrate

Here for attribute 2 and 3; Keep Up the Good Work for attribute 5; Low Priority for attribute 1; Possible Overkill for attribute 4.

6. References

[1] (2016, May 24). DOI= productlifecyclestages.com

[2] Cornell University. (2006). DOI= https://courses.cit.cornell.edu/cuttingedge/lifeCycle/03.htm

[3] Badan Pusat Statistik. (2013). DOI= www.bps.go.id

[4] Cladouhos, W. P. (1995). The multi-purpose vehicle reclassification and minivan dumping disputes bestween the united states and japan and their consistency with united states obligations under the GATT. American University International Law Review, 1109-1166.

[5] Day, G. S. (1981). The product life cycle: analysis and applications issues. Journal of Marketing Vol. 45, 60-67.

[6] Gabungan Industri Kendaraan Bermotor Indonesia (GAIKINDO). (2014). DOI= www.gaikindo.or.id

[7] Kiritsis, D., Bufardi, A., & Xirouchakis, P. (2003). Research issues on product lifecycle management and information tracking using smart embedded systems. Advanced Engineering Informatics, Vol 17, 189-202.

[8] Kotler, P. (1997). Marketing management: Analysis, planning, implementation, and control (9th ed). Upper Saddle River, NJ: Prentice Hall.

[9] Laporan Statistik Kinerja Industri Indonesia 2015. (2015). DOI= www.indoanalisis.co.id

[10] Martilla, J. A., & James, J. C. (1977). Importance-performance analysis. Journal of Marketing, 77-79.

[11] Nemcekova, M. (2008). Potentialities of product lifecycle management in automotive industry. Journal of Engineering annals of Faculty of Engineering Hunedoara, 65-70.

[12] Robinson, P., & Maher, F. (2006). Trends in automotive PLM. CPDA Annual Conference "PLM Road Map".

[13] Rogers, E. M. (1962). Diffusion of Innovations. New York: The Free Press (IV ed. 1995).

[14] Sabadka, D. (2013). Impacts of shortening product life cycle in the automotive industry. Transfer Inovacii, 251-253.

[15] Sarwono, J. (2006). Metode Penelitian Kuantitatif dan Kualitatif. Yogyakarta: Graha Ilmu.

[16] Terzi, S., Bouras, A., Garetti, M., Kiritsis, D., & Dutta, D. (2010). Product lifecycle management - from its history to its new role. International Journal of Product Lifecycle Management, 360-389.

[17] Volpato, G., & Stocchetti, A. (2008). Managing product life-cycle in the auto industry: evaluating carmakers effectiveness. International Journal of Automotive Technology and Management, 295-311.

Acknowledgements

This work is supported by Hibah PIT 9, funded by the Directorate of Research and Community Service (DRPM) Universitas Indonesia, No. NKB-0041/UN2.R3.1/HKP.05.00/2019

IOP Conf. Series: Materials Science and Engineering **831** (2020) 012021 doi:10.1088/1757-899X/831/1/012021

Localization and Mapping Algorithm for the Indoor Mobile Robot Based on LIDAR

Y Xiang[1,2], X Q Yang[1,2], W W Yang[1,2] and W H Miao[1,2]

[1] The Second Research Institute of CAAC, Chengdu, China
[2] Civil Aviation Logistics Technology Company Limited, Chengdu, China

yangxiuqing@caacetc.com

Abstract. In order to realize the autonomous Localization and navigation of indoor robots, this paper designs an indoor mobile robot positioning and mapping system based on LIDAR, and proposes a kind of localization and mapping task based on the improved RBPF-SLAM algorithm. This method mainly solves the deficiency in the original RBPF-SLAM algorithm from the aspect of improving the proposed distribution and limiting the number of resampling, making it more suitable for indoor mobile robot localization and mapping. Finally, the accuracy and rationality of the robot localization and mapping based on the RBPFSLAM algorithm based on LIDAR are verified by experiments.

1. Introduction

With the development of science and technology, the mobile robot autonomous navigation technology has developed rapidly, and now there are many kinds of navigation methods that have been widely used, such as magnetic navigation, inertial navigation, GPS navigation, laser navigation and visual navigation [1-5].

At present, the most widely used non-contact sensors in the field of robot positioning and navigation are mainly cameras and LIDAR [6,7]. The visual navigation based on the camera mainly relies on the camera to collect the environment information, through a series of image processing technology such as feature recognition matching to realize the navigation of mobile robot. The realization method is relatively simple, but is susceptible to environmental lighting, and the camera field of view is limited, so the algorithm requirements are higher. Laser navigation uses LIDAR sensors to collect distance information from the environment, realize the robot's autonomous navigation, the accuracy is relatively high, can achieve 360 degrees of information capture, and not affected by environmental lighting and other factors, can run day and night, and better night effect, therefore, the technology has been the auto pilot car. The most central sensor device in areas such as drone driving.

The LIDAR-based SLAM algorithm obtains a series of scattered point cloud data with accurate angle and distance information collected by LIDAR, then matches it by ICP (iterative nearest neighbor algorithm), and finally calculates the distance and attitude change of LIDAR relative motion by matching point cloud data at different times to complete the localization of the mobile robot itself [8,9]. Researchers at Tokyo University first used particle filter (PF) to complete SLAM in 2006, and successfully realized the application of FastSLAM in AUV navigation and positioning [10]. Because PF uses sample form instead of function form to estimate the system state and does not consider the statistical characteristics of noise, the prediction model in practical application is inaccurate. The traditional FastSLAM chooses the prior distribution as the recommended distribution [11,12], which results in the serious degradation of particles without considering the latest measurements. Resampling reduces the degree of particle degradation but results in particle dilution.

Based on previous research and the shortcomings of the traditional LIDAR localization and mapping algorithm based on particle filtering algorithm, a laser SLAM algorithm based on improving the proposed distribution and limiting the number of resampling is proposed. It mainly includes: using LIDAR to collect 3D (three-dimensional) point cloud data in environment, and obtaining the position parameters of adjacent moment robots according to ICP calculation; According to the motion model of the actual robot, the proposed distribution of particle filtering algorithm and the calculation of particle weight update map, and finally limit the number of resampling to maintain the diversity of particles makes it more realisti.

2. Indoor mobile robot positioning algorithm based on LIDAR

The mobile robot localization algorithm based on laser SLAM mainly includes: LIDAR data acquisition, relative position estimation based on ICP data matching and particle filtering. The algorithm flow chart in Figure 1.

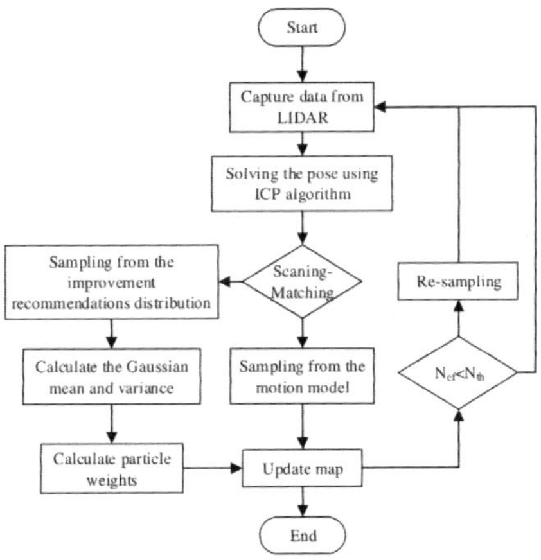

Figure 1. PF-SLAM algorithm flow chart based on LIDAR.

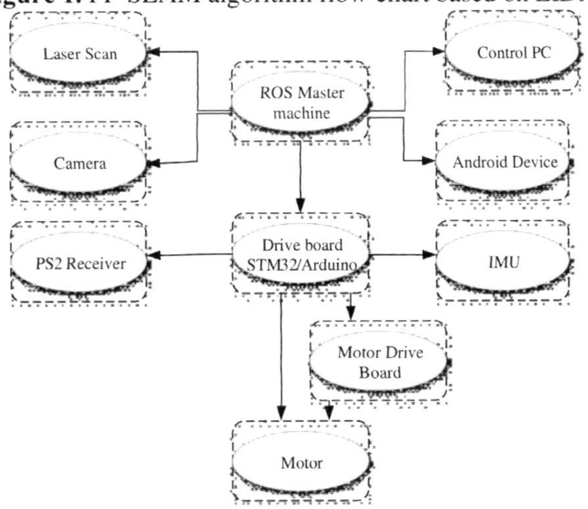

Figure 2. Pibot Mobile Robot System Block.

2.1. Laser-based mobile robot and its localization system scheme design

According to the structure of the tire, the mobile robot can be divided into track-type mobile robot and wheel-type mobile robot. Considering that the robot mentioned in this paper is carried out in indoor environment, the choice of circumference technology PIBOT specifically for ROS development of the mobile differential two-wheeled car, which is suitable for indoor flat ground walking.

The Pibot mobile robot system frame diagram shows as shown in Figure 2, removing some of the expansion peripherals, mainly consisting of four parts: mobile robot platform, LIDAR sensor, the upper machine Arduino development board and control system (ROS host).

3. Robotic LIDAR localization and mapping based on improved RBPF-SLAM algorithm

In indoor robot localization problems with particle filtering frames, each particle only needs to represent a possible attitude. But in SLAM, attitudes and maps are the status variables, so a particle needs to hold the position of a moving robot at the same time for all historical moments as well as the entire map. According to the condition Bayesian law, there are:

$$p\left(x_{1:t}, m \mid z_{1:t}, u_{1:t-1}\right) = p\left(m \mid x_{1:t}, z_{1:t}\right) \bullet p\left(x_{1:t} \mid z_{1:t}, u_{1:t-1}\right) \quad (1)$$

This decomposition actually divides SLAM into two steps of localization and mapping, which greatly reduces the complexity of SLAM problems. This treatment is called Rao-Blackwellized Particle Filtering (RBPF). The simple principle of its algorithm is to use the map and motion model of the previous moment to predict the position of the current moment, and then calculate the weight, resampling, update the particle map, loop operation [13].

However, there are two fatal problems with the above algorithm. First, the map has high requirements for the robot attitude, for any one particle, only rely on the results of motion model sampling to build a map, the error will be very large. If there were enough particles, there might be several particles that got an accurate map, but overall, the proposed distribution swayed through motion models is too dispersed and too far apart from the actual distribution.Second, frequent resampling results in particle dissipation. Each particle needs to hold the position of the moving robot at all historical moments, as well as the entire map, and frequent resampling can cause the diversity of positional postures that were long ago.

3.1. Improved PF-SLAM algorithm based on LIDAR

In view of the two major problems of the conventional PF-SLAM algorithm, based on the enlightenment of Gmapping algorithm, this paper adopts the method of improving the proposed distribution and limiting the number of resampling, can use the fewer particles to estimate the state, reduce the computational complexity, and effectively solve the phenomenon of particle degradation.

3.2. Improve the distribution of proposals

RBPF uses the odometer motion model as the proposed distribution, causing most of the particles to deviate from their real positions. In contrast to motion models, the observational model of LIDAR can give a relatively concentrated distribution. As shown in Figure 3, if the particle sampling range is changed from a flat and wide area to the peak region L represented by the LIDAR observation model, the new particle distribution can be closer to the real distribution.

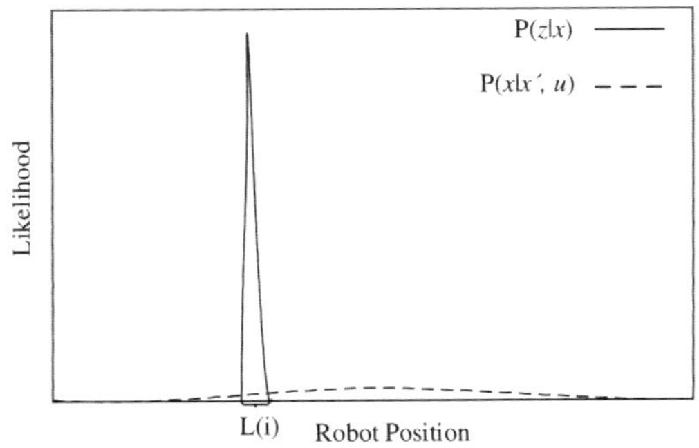

Figure 3. Change particle sampling range effect.

The scheme is to make a scan match of the prediction as the initial value after the prediction is given by the odometer motion model. The purpose of scanning matching is to find a position profile that best fits the map for the current observation. After scanning the match, we find the peak region represented by L and determine the mean and variance of the Gaussian distribution represented by the peak region. According to the literature [13], the best approximation of the true distribution is obtained by random sampling of K points in L and calculating the mean and variance according to the k-point odometer and observation model.

$$\mu_t^{(i)} = \frac{1}{\eta^{(i)}} \bullet \sum_{j=1}^{K} x_j \bullet p\left(z_t \mid m_{t-1}^{(i)}, x_j\right) \bullet p\left(x_j \mid x_{t-1}^{(i)}, u_{t-1}\right) \tag{2}$$

$$\Sigma_t^{(i)} = \frac{1}{\eta^{(i)}} \bullet \sum_{j=1}^{K} p\left(z_t \mid m_{t-1}^{(i)}, x_j\right) \bullet p\left(x_j \mid x_{t-1}^{(i)}, u_{t-1}\right) \bullet \left(x_j - \mu_t^{(i)}\right)\left(x_j - \mu_t^{(i)}\right)^T \tag{3}$$

$$\eta^{(i)} = \sum_{j=1}^{K} p\left(z_t \mid m_{t-1}^{(i)}, x_j\right) \bullet p\left(x_j \mid x_{t-1}^{(i)}, u_{t-1}\right) \tag{4}$$

Therefore, the new particles are sampled from the Gaussian distribution. Each particle is given a weight for subsequent resampling steps. The weights are calculated as follows [11]:

$$\begin{aligned} w_t^{(i)} &= w_{t-1}^{(i)} \bullet p\left(z_t \mid m_{t-1}^{(i)}, x_{t-1}^{(i)}, u_{t-1}\right) = w_{t-1}^{(i)} \bullet \int p\left(z_t \mid m_{t-1}^{(i)}, x'\right) \bullet p\left(x' \mid x_{t-1}^{(i)}, u_{t-1}\right) dx \\ &\simeq w_{t-1}^{(i)} \bullet \sum_{j=1}^{K} p\left(z_t \mid m_{t-1}^{(i)}, x_j\right) \bullet p\left(x_j \mid x_{t-1}^{(i)}, u_{t-1}\right) \end{aligned} \tag{5}$$

The upper formula is a full probability formula that takes into account the positions of K that may appear at the current moment, and calculates the weighted average of their contribution to the current observation.

3.3. Limit the number of resampling

To avoid particle dissipation, the number of resampling should be limited. The diversity and accuracy of particles are maintained at a high level as the robot continues to move towards unknown areas. Particles in the local region maintain a high accuracy, but the cumulative error is always superimposed, if frequent resampling, the diversity of particles will disappear, the longer-term position attitude will become more and more single.

The purpose of resampling is to enhance particles close to the real value and discard particles that are far from the true value. Prior to the loop, even though some particles were far removed from the true value, existing observations were not sufficient to distinguish the correct particles from the wrong particles, so resampling was meaningless at this time. Resampling can only be as effective as it should be if the new observation completes completely widen the weight gap between the correct and the wrong particles after the loop has occurred.

Therefore, it is necessary to determine whether a loop back ring has occurred, as mentioned in the literature [13], and it can be evaluated below to determine whether the loop back ring occurs.

$$N_{eff} = \frac{1}{\sum_{i=1}^{N}\left(\tilde{w}^{(i)}\right)^2} \tag{6}$$

If N_{eff} is larger, the smaller the particle weight gap. When N_{eff} is lowered below a certain threshold, it indicates that there is a large gap between the distribution of particles and the true distribution, showing at the particle level that some particles are very close to the true value, and many particles are far from the true value. This is exactly what happens when the loop is occurring, so choose this time for resampling.

4. Experimental results and analysis

In order to verify the feasibility and measurement accuracy of the laser-based ground robot localization and mapping algorithm proposed in this paper, this paper is carried out through the robot localization and mapping experiment and the robot navigation experiment.

4.1. Robot localization and mapping experiments

In order to verify the accuracy of the robot localization and mapping algorithm proposed in this paper, the proposed algorithm is run on the ROS robot operating system, and the specific steps include:

1) Add LIDAR drive, using Flash LIDAR in this experiment, you can find the corresponding LIDAR drive from the ROS open source Package, and modify the corresponding parameters, baud rate, interface, coordinate system, etc.

2) Add robot start-up node: the mobile robot platform at the time of this experiment has an Arduino development board, which has the Arduino node, connects the mobile robot platform, and completes the PID control parameter setting of the mobile platform, encoder data reading and conversion, communication serial setting.

3) Add the slam-mapping node: Add the node and configure the mapping parameter, where the key parameters are shown in Table 1.

Table 1 The parameters set in the PF-SLAM system

Name	Value	Description
Xmin, ymin, xmax, ymax	/	Initialize the map size
particales	8	Number of particles
resampleThreshold	0.5	Resampling threshold
iterations	5	Number of iterations of the ICP algorithm
mininumScore	280	Scan match threshold
map_update_interval	0.01	Map update frequency

The environment selected in this experiment is a laboratory, as shown in Figure 4, which shows that the environment is relatively complex and somewhat narrow. According to the map established in Figure 4, we can see that:

1) The algorithm proposed in this paper can accurately establish the environment map. This shows that the positioning algorithm proposed in this paper supports real-time viewing and displaying the position of motion robots.

2) Comparing with the initial position result of moving robot and the position of the actual robot, The

feasibility of positioning and drawing algorithms proposed in this paper is verified.

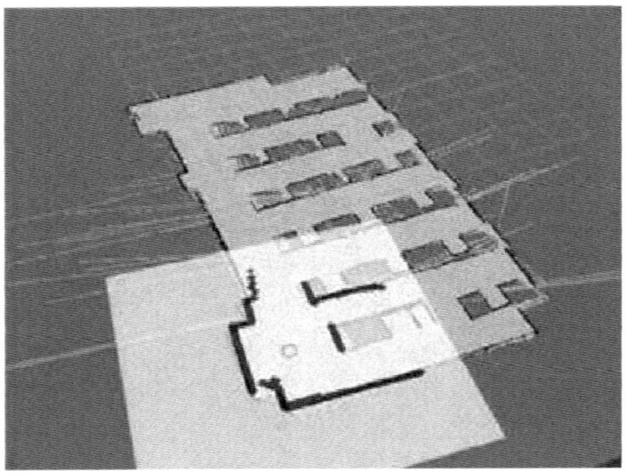

Figure 4. Rviz view of the initial location of the moving robot

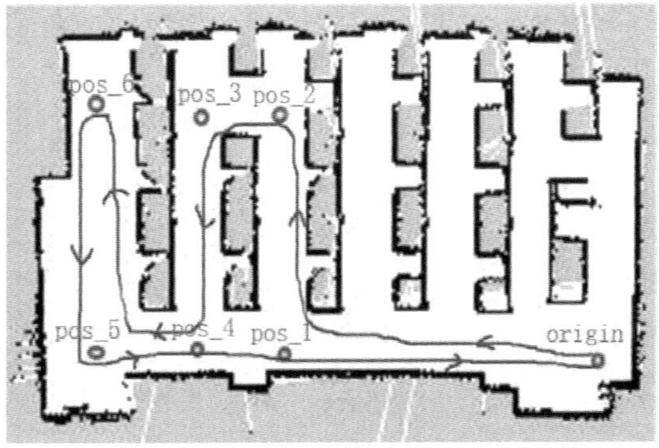

Figure 5. Set five position points and robot motion Trajectories

4.2. Robot navigation experiment

In order to verify the accuracy of the indoor robot navigation algorithm proposed in this paper, this section first uses remote control to control the robot's map of building indoor environment, and then simulates the motion of the moving robot on the map, setting six position points in turn, as shown in Figure 5, the red circle represents the position of the robot. The blue arrow represents the direction of the robot's motion. After completing the localization, the robot uses the algorithm proposed in this paper to independently plan the path to the specified five positions and finally return to the starting point according to the path planning and navigation algorithm in the literature [14].

Running navigation algorithms on robots, mobile robots begin to navigate autonomously, and mobile robots navigate the actual environment as shown in Figure 6. Similarly, the red arrow points in the direction of the robot's motion, which returns to its starting position after a circle of movement along the planned path.

Figure 6. The actual movement of robots in indoor environments

Figure 7 is a view of the Rviz that the robot moves. The red circle in the figure represents the robot, and the blue arrow represents the direction of the robot's motion. The robot starts from the starting point and moves one lap along the predetermined route back to the starting point.

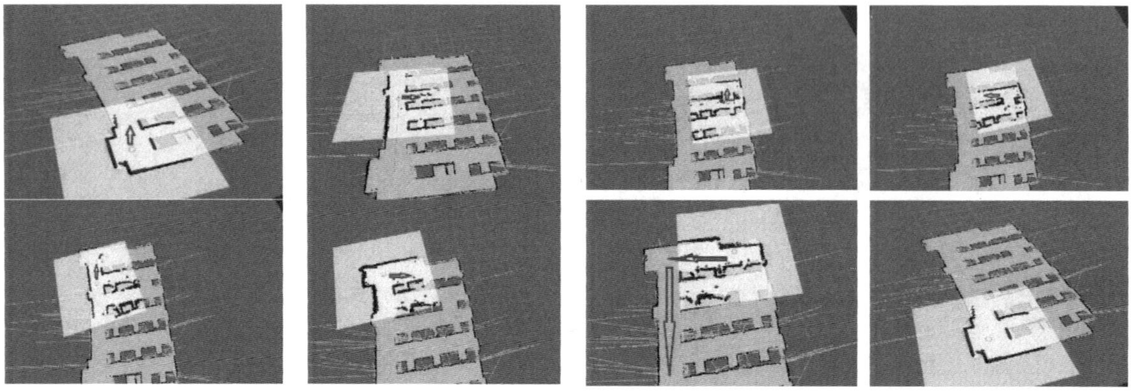

Figure 7. Rviz view of robots planning path movements

From Figures 6 and 7, it can be seen that:
1) the mobile robot can plan accurate autonomous navigation to the set six position points according to the positioning algorithm proposed in this paper. It is explained that the localization algorithm proposed in this paper has good localization accuracy and can be applied well to navigation applications in complex indoor environment.
2) Qualitatively, the motion trajectory is biased from the design trajectory, mainly because the laboratory environment is complex, and the algorithm in this paper does not consider the loop-back detection process.
In summary, the RBPF-SLAM improves the standard in two aspects: suppressing particle degradation and improving particle depletion. It improves the accuracy of navigation, positioning and feature position estimation. It is of great significance for long-range navigation and concealment tasks.

5. Conclusions

In order to realize the autonomous positioning and navigation of indoor robots, LIDAR is used as a data acquisition sensor, an all-in-one mobile robot localization and mapping system is designed and constructed, and an improved laser SLAM algorithm is proposed in combination with the traditional RBPF-SLAM algorithm according to the complex characteristics of indoor environment. The feasibility and accuracy of the algorithm proposed in this paper are verified by localization and mapping experiments and robot navigation experiments.

6. References

[1] Pappone, Carlo, et al. Robotic magnetic navigation for atrial fibrillation ablation. Journal of the American College of Cardiology 47(07) : 1390-1400, 2006.

[2] Bortz, John E. A new mathematical formulation for strapdown inertial navigation. IEEE transactions on aerospace and electronic systems 1 (1971): 61-66.

[3] MADISON, R., et al. Vision-aided navigation for small UAVs in GPS-challenged environments. En AIAA Infotech@ Aerospace 2007 Conference and Exhibit. 2007: 2986.

[4] Liu, Pei, Jun Chen, and Mingying Zhang. Automatic control system of orchard tractor based on laser navigation. Transactions of the Chinese Society of Agricultural Engineering27(03) : 196-199, 2011.

[5] Bonin-Font, Francisco, Alberto Ortiz, and Gabriel Oliver. Visual navigation for mobile robots: A survey. Journal of intelligent and robotic systems 53(03): : 263-296, 2008.

[6] Pershina, J. S., S. Ya Kazdorf, and A. V. Lopota. Methods of Mobile Robot Visual Navigation and Environment Mapping. Optoelectronics, Instrumentation and Data Processing 55(02) 181-188,

2019.

[7] MENG, Xiangrui, et al. A Navigation Framework for Mobile Robots with 3D LiDAR and Monocular Camera. En IECON 2018-44th Annual Conference of the IEEE Industrial Electronics Society. IEEE, 2018: 3147-3152.

[8] SONG, Kai-Tai, et al. Navigation Control Design of a Mobile Robot by Integrating Obstacle Avoidance and LiDAR SLAM. En 2018 IEEE International Conference on Systems, Man, and Cybernetics (SMC). IEEE, 2018:1833-1838.

[9] CHENG, Yi; WANG, Gong Ye. Mobile robot navigation based on lidar. En 2018 Chinese Control and Decision Conference (CCDC). IEEE, 2018:1243-1246.

[10]Maki, T.; Kondo, H.; Ura, T.; Sakamaki, T. Photo Mosaicing of Tagiri Shallow Vent Area by the AUV "Tri-Dog 1" using a SLAM based Navigation Scheme. In proceedings of the Oceans MTS/ IEEE, Boston, MA, USA. 18-21 September 2006:78-85.

[11]Lee, S.H.; Eoh, G.; Lee, B.H. Relational FastSLAM: an improved Rao-Blackwellized particle filtering framework using particle swarm characteristics. Robotica 2016, 34, 1282-196.

[12]Yokozuka, M.; Matsumoto, O. Sub-map dividing and realignment FastSLAM by blocking gibbs MCEM for large-scale 3-D grid mapping. Adv. Robotics 2012(26), 1649-1675, 2012.

[13]YATIM, Norhidayah Mohamad; BUNIYAMIN, Norlida. Development of Rao-Blackwellized Particle Filter (RBPF) SLAM algorithm using low proximity infrared sensors. En 9th International Conference on Robotic, Vision, Signal Processing and Power Applications. Springer, Singapore, 2017: 395-405.

[14]Mohanty, Prases K., and Dayal R. Parhi. Optimal path planning for a mobile robot using cuckoo search algorithm. Journal of Experimental & Theoretical Artificial Intelligence 28(02): 35-52, 2016.

7. Acknowledgements

The authors appreciate the supports of the National Natural Science Foundation of China— CAAC joint research fund for civil aviation (No. U1933123).

Pose Calibration for 2D Laser Profiler Integrated in Five-Axis Machine Tools

S Zhang[1], X Duan[2], C Zeng[1], K Shi[2], X Zhang[2], Q Bi[1] and Y Wang[1]

[1] State Key Laboratory of Mechanical System and Vibration, School of Mechanical Engineering, Shanghai Jiao Tong University, Shanghai 200240, P.R. China
[2] AVIC Xi'an Aircraft Industry (Group) Co., Ltd., Xi'an 710089, Shaanxi, P.R. China

biqz@sjtu.edu.cn

Abstract. Laser scanning is a fast and effective method to acquire the 3D surface profiles of part. To implement on-machine measurement (OMM), the 2D laser profiler could be integrated in five-axis machine tools and moves under the command of the computer numerical control (CNC) system. The 2D profile data and the corresponding machine coordinates are acquired and saved by the CNC system, and then converted to 3D point cloud using the kinematic model. Therefore, calibration of the position and orientation of the laser profiler with respect to the machine tool will directly influence the inspection accuracy. In this paper, a pose calibration method is proposed to identify the position and orientation of the laser sensor. A laser scanning system integrated in five-axis machine tool is developed. The position and orientation deviations between the local laser coordinate system (LCS) and tool coordinate system (TCS) are defined. Then, a designed calibration block is scanned, and its feature is reconstructed. Furthermore, the position and orientation of the laser profiler is identified by comparing the reconstructed features with the designed features. Finally, the position and orientation deviations parallel to five axes of the machine tool are compensated by coordinate offset, and the remaining orientation deviation is corrected by the kinematic model during converting measurement data to 3D point cloud. Experiment is carried out to validate the proposed method.

1. Introduction

Laser scanning is a non-contact measuring method with high speed and sufficient accuracy[1], which is widely used for microscopy[2], map building[3, 4], shape acquisition and quality assessment[5]. Bi [6] adopted a laser scanner mounted on the spindle to measure the deformed large skin. Liu [7] employed an iso-planar-based on-machine scanning method to obtain the shape of complex surface. Sharifzadeh [8] used laser scanners integrated in robot to detect the small abnormalities on large complex free-form surfaces. However, laser scanning is usually not as accurate as coordinate measurement machine (CMM) with a contact-type probe [9].

A lot of research is devoted to improving the accuracy of laser scanning. Isheil [9] studied the effects of three geometrical parameters of sensor positioning, and built a global and local correction model to reduce the mean errors and standard deviations, respectively. Li [10] extended the depth-of view of the portable 3D scanning system from the Scheimpflug condition by tilting the charge coupled device (CCD) plane, and improved the measurement accuracy via segmental calibration. Sun [11] proposed a vision measurement model and a corresponding calibration method for the laser displacement sensor to avoid linear approximation and be implemented conveniently. Genta [12] proposed a thorough approach based on a reference ball plate to calibrate the triangular laser scanner. Abu-Nabah [13]

IOP Conf. Series: Materials Science and Engineering **831** (2020) 012022 doi:10.1088/1757-899X/831/1/012022

presented a simplified method for laser vision sensor calibration using a rectangular notch calibration block with high precision. Most above calibration methods are proposed for scanner uncertainties, which could be calibrated and compensated by the manufacturer [9]. However, in practical applications, laser scanning system is implemented by mounting a commercial laser sensor at the end of a machine tool or robot. The profile information and the corresponding machine coordinates are sampled at the same time and then converted to 3D points through kinematic model [14]. During the installation of the laser sensor, assembly error is inevitable. Therefore, in addition to the performance of the laser sensor, its assembly error affects the scanning accuracy. The true installation position and posture of the sensor need to be accurately calibrated to improve measurement accuracy. However, there is a lack of methods for calibrating the pose of the laser scanner at present.

This work proposes a pose calibration method for the 2D laser profiler integrated in five-axis machine tool. The laser scanning system is introduced briefly in Section 2. Then, Section 3 defines position the and orientation deviations and proposes a pose calibration method based on a designed calibration block. Experiment is conducted in Section 4, and conclusions are drawn in Section 5.

2. Laser scanning system

The employed experiment platform is a dual-five-axis mirror milling system, which is composed of a five-axis milling head and a five-axis support head, as shown in Figure 1. The milling head consists of three translations (following X_1, Y_1 and Z_1) and two rotations (following A_1, C_1), the support head also consists of three translations (following X_2, Y_2 and Z_2) and two rotations (following A_2, B_2). A 2D laser profiler (Keyence LJ-V7200) mounted on the support head moves under the command of CNC system to implement on-machine scanning. The CNC system acquires the profile data from laser sensor and the corresponding machine coordinates from machine tool at the same time and saves it to the measurement file. The specifications of the adopted profiler are listed in Table 1.

Figure 1. Experiment setup.

Table 1. Specifications of the 2D laser profiler.

Height (Z)	Parameters	Width (X)	Parameters
Reference distance	200 mm	Profile data count	800 items
Measurement range	±48 mm (F.S.=96 mm)	Measurement range	62 mm
Linearity	±0.1% of F.S.	Profile Data interval	100 μm
Repeatability	1 μm	Repeatability	20 μm

3. Calibration method

3.1. Position and orientation deviation description

LCS ($O_L X_L Y_L Z_L$) is a local measurement coordinate system fixed with the reference point of the laser profiler, and TCS ($O_{T2} X_{T2} Y_{T2} Z_{T2}$) is fixed with the tip point of support head. Limited to the structure of support head and assembly errors of the laser profiler, LCS deviates from TCS, as shown in Figure 2(a). The positional deviations are the coordinates of LCS origin in TCS, defined as $(\delta x, \delta y, \delta z)$. The

IOP Conf. Series: Materials Science and Engineering **831** (2020) 012022 doi:10.1088/1757-899X/831/1/012022

orientation deviations are the posture differences between LCS and TCS, represented by (α, β, γ), as shown in Figure 2(b). α is the angle between Z_{T2} and the projection of the Z_L axis in the $Y_{T2}Z_{T2}$ plane. β is the angle between X_{T2} and the projection of the X_L axis in the $Z_{T2}X_{T2}$ plane. γ is the angle between X_{T2} and the projection of the X_L axis in the $X_{T2}Y_{T2}$ plane.

The target of the pose calibration for 2D laser profiler is to identify and correct α, β, γ, δx, δy and δz. The calibrated deviations α, β, δx, δy and δz, which are parallel to the axes of support head, could be adjusted by setting machine zero point of support head. After adjustment, the central axis of the laser plane Z_L and origin of LCS coincide with TCS, as shown in Figure 2(c). The remaining orientation deviation γ can be compensated when the measurement data is converted to point cloud. Assume that the measured height of the *i-th* measuring point $\boldsymbol{P}_i$ is h in the LCS, then the $\boldsymbol{P}_i$ can be represented as:

$$ {}^{L}P_i = \left[{}^{L}x_{P_i}, {}^{L}y_{P_i}, {}^{L}y_{P_i} \right]^T = \left[\Delta d(i-400), 0, h \right]^T \tag{1} $$

where $\Delta d = 0.1mm$ is the profile data interval of the laser profiler. As illustrated in Figure 2(d), convert ${}^{L}\boldsymbol{P}_i$ to ${}^{T_2}\boldsymbol{P}_i$ considering orientation deviation γ:

$$ {}^{T_2}\mathbf{P}_i = \left[{}^{T_2}x_{P_i}, {}^{T_2}y_{P_i}, {}^{T_2}z_{P_i} \right]^T = \left[\Delta d(i-400)\cos\gamma, \Delta d(i-400)\sin\gamma, h \right]^T \tag{2} $$

${}^{T_2}\mathbf{P}_i$ can be further converted to ${}^{W_2}\mathbf{P}_i$ using the kinematic transformation model proposed in Ref. [15] to obtain 3D point cloud in the workpiece coordinate system.

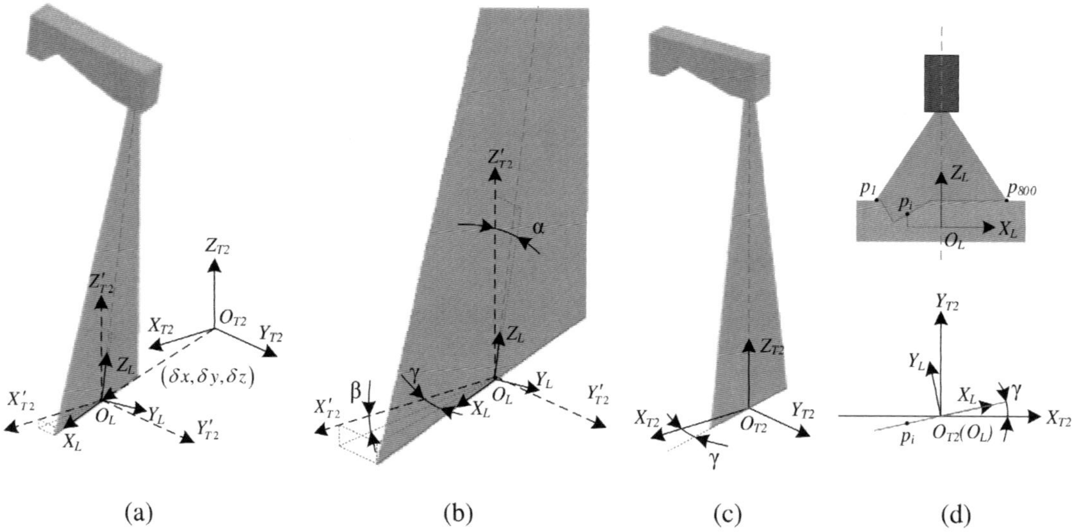

Figure 2. Pose deviation definition and correction: (a) Position deviations; (b) Orientation deviations; (c) Correction of α, β, δx, δy and δz; (d) Correction of γ.

3.2. Calibration model of β

When orientation deviation β is nonzero, the measured profile will deviate from the reference profile, which is parallel to X_2Y_2 plane, as shown in Figure 3(a). If B_2 axis rotates, β changes accordingly. When the recognized β is zero, the angle of B_2 currently is the calibration value of β.

3.3. Calibration model of α

As shown in Figure 3(b), O is the pivot of A_2 axis, A is the laser source, B is the image point of the laser on the workpiece, C is the foot point of O on the reference plane, $\overline{OC} = l$, $\overline{OA} = n$, $\angle COA = \sigma$. When A_2 axis rotate φ, A moves to A', B moves to B', and measured distance h changes accordingly. φ and h satisfies the following equation:

$$ h = \frac{l - n\cos(\sigma - \varphi)}{\cos(\alpha - \varphi)} \tag{3} $$

l, n, σ and α are the parameters to be determined. The curve about φ and h can be obtained through scanning a reference plane by rotating the A_2 axis. Fit the measured curve using the least square method to identify l, n, σ and α.

3.4. Calibration model of γ, δx, δy and δz

After correction of α and β, Z_L is parallel to Z_{T2}, but laser plane may be not coincided with $X_{T2}Z_{T2}$ plane. Remaining γ, δx, δy and δz can be calibrated by scanning a reference hole. When the laser profiler scanning the hole along Y_2 direction, the laser image line is not perpendicular to the scanning direction due to the orientation deviation γ, as shown in Figure 3(c). The scanned point on the edge of the hole can be described as the following parametric equation:

$$\begin{cases} x - x_0 = r\cos\theta \\ y - y_0 = r\sin\theta \end{cases} \tag{4}$$

The corresponding recognized distance along the laser image line and along the scanning direction are

$$\begin{cases} dx = x' - x_0 = r\dfrac{\cos\theta}{\cos\gamma} \\ dy = y' - y_0 = r\sin\theta - r\cos\theta\tan\gamma \end{cases} \tag{5}$$

Eliminate θ, we get a new curve equation

$$r^2 = \left(x' - x_0\right)^2 + \left(y' - y_0\right)^2 + 2\left(x' - x_0\right)\left(y' - y_0\right)\sin\left(\gamma\right) \tag{6}$$

The right schematic in figure 3(c) illustrates the curve described by Eq.(6), which is the recognized hole contour without compensation for γ. The recognized contour is fitted using least squares to identify γ, x_0, y_0. Fit the plane where the hole is located and get its z coordinate z_0. Comparing (x_0, y_0, z_0) to the center coordinates of reference hole yields $(\delta x, \delta y, \delta z)$.

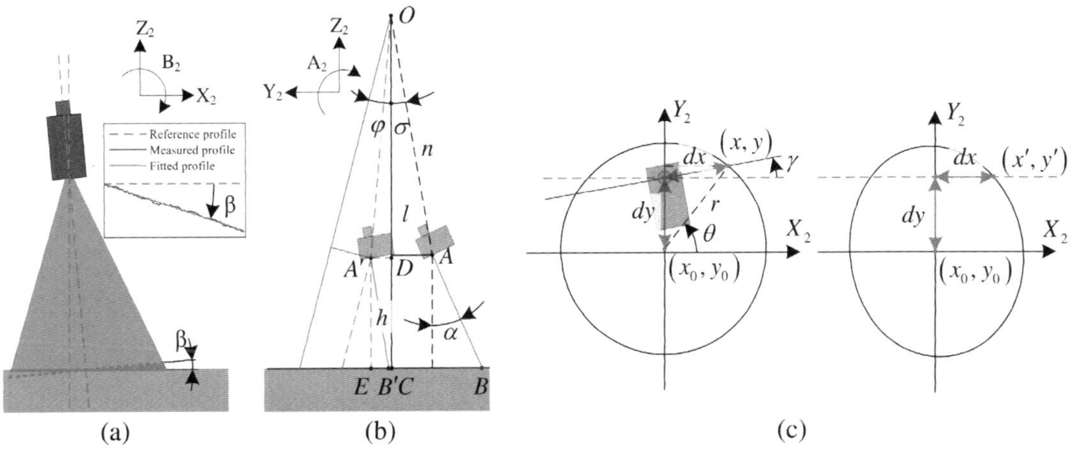

Figure 3. Calibration model: (a) Model of β; (b) Model of α; (c) Model of γ, δx, δy and δz.

4. Experiment

A designed calibration block with a hole, as shown in Figure 4, is utilized to calibrate the pose of the laser profiler. The bottom of the central notch and the side wall of the hole are finished as the reference plane and hole. Its exact dimensions are measured by CMM and marked in Figure 4 (a). The block is mounted on the spindle of the milling head by a shank, and its edge is adjusted to be parallel to the machine coordinate system via spindle positioning. Therefore, at the initial pose of the milling head, the reference plane is parallel to XY plane, and the centre coordinates of the reference hole is (0.043,-27.470, 0.000).

The calibration experiment is carried out in four steps:

Step1: Calibration of β. Scan the reference plane by rotating B_2 axis while $A_2 = 0°$ and obtain the curve about β and B_2, as shown in Figure 5(a). When recognized β is 0°, B_2 is 0.277°. So $\beta = 0.277°$.

Step2: Calibration of α. Scan the reference plane by rotating A_2 axis while $B_2 = 0.277°$ and obtain the curve about h and φ, as shown in Figure 5(b). Fit the curve using Eq.(3) and obtain $\alpha = -0.899°$.

Step3: Calibration of γ, δx, δy and δz. Scan the reference hole along Y_2 direction while $A_2 = -0.899°$ and $B_2 = 0.277°$ and recognize the contour of the hole, as shown in Figure 5(c). Fit the measured edge using Eq.(6) and obtain $\gamma = 0.227°$, $x_0 = -129.830mm$, $y_0 = -95.046mm$, $z_0 = 112.748mm$. Corresponding position deviations are $\delta x = 129.873mm$, $\delta y = 67.576mm$, $\delta z = -112.748mm$.

Step4: Correction of deviations. Set $A_2 = -0.899°$, $B_2 = 0.277°$, $X_2 = 129.873mm$, $Y_2 = 67.576mm$, $Z_2 = -112.748mm$ as the new zero point of support head to correct α, β, δx, δy and δz. Use Eq.(2) to compensate for γ during measurement data converting.

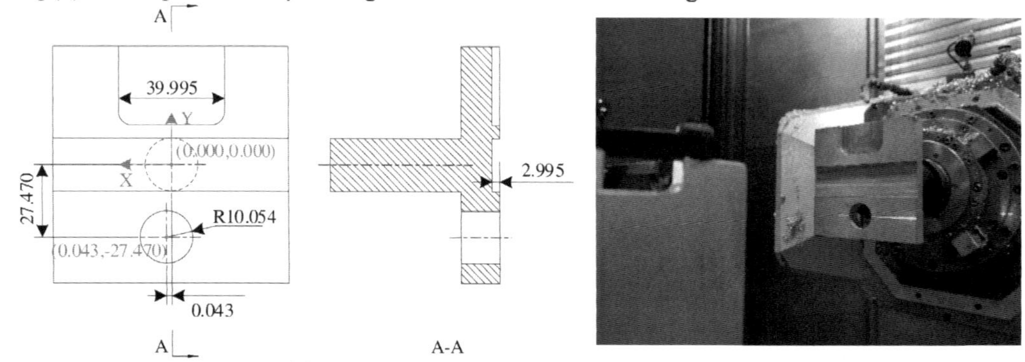

Figure 4. Calibration block: (a) Dimensions of block (unit: mm); (b) Photo of block.

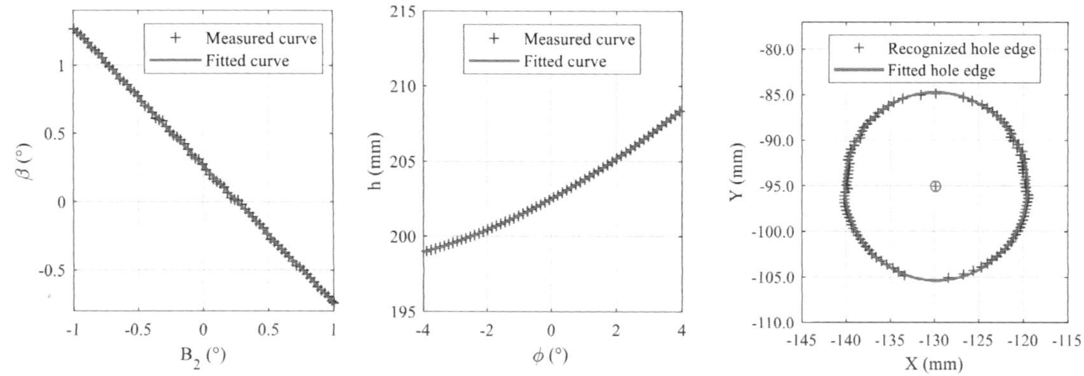

Figure 5. Experiment results: (a) Calibration of β; (b) Calibration of α; (c) Calibration of γ, δx, δy, δz.

The calibration results can be used to correct the assembly deviations. To verify the effectiveness of the proposed calibration method, the reference hole was scanned with and without correcting the deviations. The scanned hole contours are then recognized and then fitted to ellipses, and the lengths of semi-major and semi-minor axis are listed in Table 2. Due to orientation deviations, the uncorrected hole contour, which is the projection of a circle on the inclined plane, will appear elliptical. The difference between the semi-major and semi-minor axis characterizes the proximity of the recognized hole contour to the true hole contour, which is a circle. In other words, the smaller the difference, the higher the scanning and recognition accuracy. Obviously, the fitted ellipse with correction has a smaller difference, indicating that the measurement accuracy is improved after correction.

Table 2. Ellipse fitting results.

	Semi-major axis (mm)	Semi-minor axis (mm)	Difference (mm)
With correction	10.13	10.12	0.01
Without correction	10.15	10.09	0.06

5. Conclusion

This paper presents a pose calibration method for 2D laser profiler integrated in five-axis machine tool. Three angular deviations and three translational deviations are defined to represent the orientation and position of the laser profiler. Then, a calibration method is proposed to identify and correct the defined deviations using a designed calibration block. Experiment demonstrates the feasibility of the approach. The proposed calibration method makes full use of the large numbers of measured points through curve fitting, reduces the random error and improves measurement accuracy.

6. References

[1] Li Y and Gu P 2004 Free-form surface inspection techniques state of the art review *Comput. Aided D.* **36** 1395-1417
[2] Nadolny K and Kaplonek W 2012 Confocal laser scanning microscopy for characterisation of surface microdiscontinuities of vitrified bonded abrasive tools *International Journal of Mechanical Engineering Robotics Research* **1** 14-29
[3] Kita S, Lee J H and Okamoto S 2016 3D Map Building Using Mobile Robot with Scan Device *International Journal of Mechanical Engineering Robotics Research* **5** 191-195
[4] Yoo W, Lee S-H and Lee B H 2016 An Enhanced Map Building Framework Based on the Scan Similarity *International Journal of Mechanical Engineering Robotics Research* **5** 39
[5] Son S, Park H and Lee K H 2002 Automated laser scanning system for reverse engineering and inspection *Int. J. Mach. Tools Manuf.* **42** 889-897
[6] Bi Q, Huang N, Zhang S, Shuai C and Wang Y 2018 Adaptive machining for curved contour on deformed large skin based on on-machine measurement and isometric mapping *Int. J. Mach. Tools Manuf.* **136** 34-44
[7] Liu H, Wang Y, Huang X and Xue L 2013 Isoplanar-based adaptive sampling for model-unknown sculptured surface coordinate metrology using non-contact probe *Int. J. Adv. Manuf. Technol.* **64** 1695-1707
[8] Sharifzadeh S, Biro I, Lohse N and Kinnell P 2018 Abnormality detection strategies for surface inspection using robot mounted laser scanners *Mechatronics* **51** 59-74
[9] Isheil A, Gonnet J P, Joannic D and Fontaine J F 2011 Systematic error correction of a 3D laser scanning measurement device *Opt. Lasers Eng.* **49** 16-24
[10] Li J, Guo Y, Zhu J, Lin X, Xin Y, Duan K and Tang Q 2007 Large depth-of-view portable three-dimensional laser scanner and its segmental calibration for robot vision *Opt. Lasers Eng.* **45** 1077-1087
[11] Sun J, Zhang J, Liu Z and Zhang G 2013 A vision measurement model of laser displacement sensor and its calibration method *Opt. Lasers Eng.* **51** 1344-1352
[12] Genta G, Minetola P and Barbato G 2016 Calibration procedure for a laser triangulation scanner with uncertainty evaluation *Opt. Lasers Eng.* **86** 11-19
[13] Abu-Nabah B A, ElSoussi A O and Al Alami A E K 2016 Simple laser vision sensor calibration for surface profiling applications *Opt. Lasers Eng.* **84** 51-61
[14] Liu H, Wang Y, Jia Z and Guo D 2015 Integration strategy of on-machine measurement (OMM) and numerical control (NC) machining for the large thin-walled parts with surface correlative constraint *Int. J. Adv. Manuf. Technol.* **80** 1721-1731
[15] Zhang S, Bi Q, Ji Y and Wang Y 2019 Real-time thickness compensation in mirror milling based on modified Smith predictor and disturbance observer *Int. J. Mach. Tools Manuf.* **144** 103427

IOP Conf. Series: Materials Science and Engineering **831** (2020) 012022 doi:10.1088/1757-899X/831/1/012022

Acknowledgments
The authors gratefully acknowledge the financial support of the National Natural Science Foundation of China (Grant No. 51875357), the Advanced Technology Joint Research Fund of Shanghai Academy of Spaceflight Technology (USCAST2019-16) and the Joint Innovation Fund of China Academy of Launch Vehicle Technology (CALT201809).

IOP Conf. Series: Materials Science and Engineering **831** (2020) 012023 doi:10.1088/1757-899X/831/1/012023

Damage state model for fatigue of CFRP

Amir Siddiqui, Asim Tewari and Anirban Guha

Department of Mechanical Engineering, IIT Bombay, Powai, Mumbai 400 076, India

asim.tewari@iitb.ac.in

Abstract. Fatigue is the major failure mode for any structure under cyclic loading. In particular, composite materials exhibit highly complex damage mechanisms and are often overdesigned due to lack of understanding under cyclic loading. Many papers have been written on the modelling of fatigue in fiber reinforced composites. All such models they assumed some internal damage parameter which themselves defined by the mechanical property. However, the growth of these damage parameters are themselves defined by the mechanical property degradation. Hence these fatigue models do not give actual stiffness degradation. In this study, actual 3D microstructural damage in CFRP under cyclic loading was quantitatively measured. Damage micromechanisms viz., fiber breakage, matrix microcracking, debonding, and fiber pull-out were estimated and a model to predict the property degradation with these damage mechanisms (parameters) was developed. Thus, the property degradation in CFRP under cyclic loading was modelled with independently measured damage parameters.

1. Introduction

Carbon Fiber-reinforced composites are increasingly used as structural parts in the fields of aerospace, aviation and transport vehicles for their excellent specific strength and stiffness. Most of these structures are built to be in service for a long time during which they will be subjected to fatigue loads. The strength and stiffness degradation of the composite under fatigue loading is not well understood, which leads to conservative and overdesign of composite structure. Therefore it is necessary to understand how strength and stiffness are degrading under cyclic loading. Many papers have been written on the modelling of fatigue in fiber reinforced composites [1-4]. Degrieck and Van [1] have classified fatigue modelling in three categories as fatigue life models, progressive models, and Phenomenological models. In these models, they assumed some internal damage parameter which themselves defined by the mechanical property. Therefore these fatigue models do not give actual stiffness degradation.

In this paper, actual 3D microstructural damage such as fiber crack, fiber pull-out, fiber/matrix debonding and matrix crack were quantitatively measured. The damage induced due to fatigue loading of unidirectional carbon fiber reinforced polymer (CFRP) composites was quantified using optical microscopy. Finite element analysis (FEA) was used to find effective stiffness of damaged composite and undamaged composite. Quantitatively measured damages were used as input for FEA. Stiffness obtained using FE analysis was compared with experimental results.

2. Experimental Work

An organized experimental study was performed, consisting of tensile and fatigue tests of unidirectional coupons [5]. The coupons were made from AS4 carbon prepreg (carbon fiber)/914C (epoxy resin) with 0.6 volume fraction. As per ASTM 3039, test specimens were prepared, and bidirectional glass/epoxy tabs were glued at their ends. The length and width of the coupons were 250 mm and 15 mm, respectively.

Their nominal thickness was 2.4 mm. The length of the tabs with a thickness of 2.5 mm was 70 mm leaving a gauge length of 110 mm. Static tests were performed under displacement control at a speed of 2 mm/min. Fatigue testing were load control with different stress ratios, R = 0.5 and R = 0.7 (T–T). Interrupted cyclic testing was performed to obtain a different degree of internal damages. Coupons with varying degrees of internal damage were sliced, mounted, and polished for characterization under an optical microscope to quantify damages.

Damage measurement was done by using quantitative stereology. Stereology [6] is a method in which two-dimensional (2D) measurements can obtain quantitative data about three-dimensional (3D) parameters defining the structure from micrograph images. Fiber cracks were calculated in terms of fiber cracks per unit volume, $(N_v)_f$. It is defined by,

$$(N_v)_f = \frac{(N_A)_f}{H} \quad (1)$$

Where $(N_A)_f$ is the number of fiber cracks per unit area in μm^{-2} and H is the mean caliper diameter in μm. Fiber/matrix debonding and matrix crack were estimated by measuring surface area per unit volume of debonding and matrix crack, given by the stereological equation:

$$S_v = 2I_L \quad (2)$$

Where S_v is the surface area per unit volume in μm^{-1} and I_L is the average number of intersection points per unit length of a test line in μm.

3. Finite element model generation

The finite element analysis usually involves a study of a representative volume element (RVE) of a fiber with a periodic packing sequence. Fiber distribution at the cross-section of the composite lamina is usually quite random [7]. For simplicity, the periodic arrangement of fibers in the micromechanics model was assumed so that RVE can be isolated. The most commonly used periodic sequences of fibers are a square array and hexagonal array, as shown in figure 1 (a)-(b). In this study, we used RVE in which fibers are arranged in hexagonal array and load is applied in the fiber direction, as shown in figure 1(c).

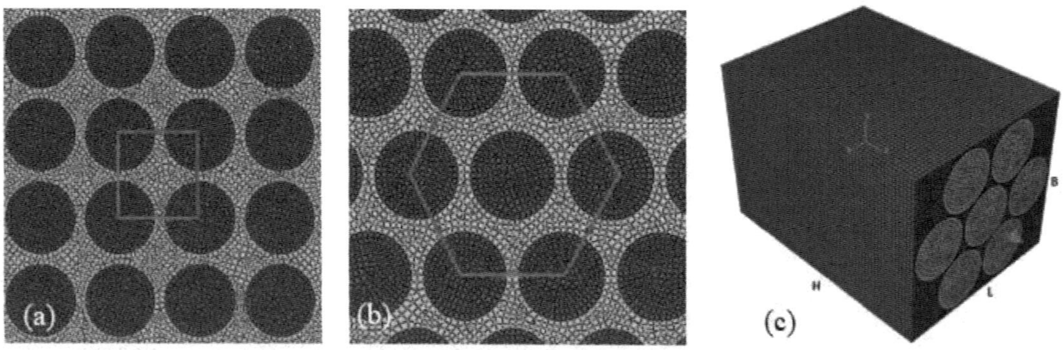

Figure 1. Fiber arrangement in RVE (a) Square array, (b) Hexagonal array (c) Hexagonal fiber array with loading along the fiber direction (L, B and H in μm)

For this study fibers were arranged in hexagonal pattern on the rectangle such that they did not intersect any boundary of rectangle. The RVE consisted of three phases, namely polymer matrix, carbon fiber and fiber/matrix interface. Elastic model were used for the matrix and fiber/matrix interface. The fiber has a higher modulus along the longitudinal direction compared to the transverse direction. Hence, carbon fibers were modelled as elastic transversely isotropic. Material properties for fiber E_L = 225 GPa, E_T = 15 GPa, ν =0.2, G_{LT} = 15 GPa, G_{TT} = 7 GPa and for epoxy E = 4 GPa, ν =0.35, X_T = 75 MPa, X_C = 150MPa and S = 70 MPa [8]. Cylindrical fibers were considered with a diameter of 7 um. The interface was modelled as hollow cylinder, around the fiber, with thickness of 0.1 μm. Fiber volume fraction of 0.6 was taken as measured from the specimen [5]. Fiber crack were introduced by assigning very low stiffness value to a row of elements in fiber.

Periodic boundary conditions were applied by using constraint equation to RVE from Barbero [9]. The degree of freedom of each pair of nodes of opposite vertices, edges, and faces of the RVE kinematic constraints are applied. The degree of freedom of nodes and applied strains are variable in these equations. Depending on which position the nodes are -edges, faces, or vertices - a different set of equations must be used to their degrees of freedom to solve compatibility issues between various kinematic constraints.

4. Results and Discussion

The tensile strength and stiffness of unidirectional CFRP were experimentally measured as 1855.3 ± 125.0 MPa and 137.5 ± 4.5GPa, respectively. Fatigue testing (till fracture) of coupons was carried out with different load factors. S-N curve was plotted in terms of maximum stress (σ_{max}) on the ordinate (Y-axis) and the number of cycles to failure (N_f) on the abscissa (X-axis) with semi-log scale. S-N curve obtained from fatigue test for samples loaded along the direction of fiber (referred as 0^0) is shown in figure 2. The endurance limit for 0^0 specimen was about 1480 MPa i.e. the specimen could sustain more than 1 million load cycles without failure.

The damage progression in the test specimen was measured with the help of optical microscopy and stereology. Optical microscope was used to obtain the micrographs of different samples. Figure 3(a), (b) displays micrographs of 0^0 specimen. Fiber cracks, fiber missing portion and fiber/matrix interface debonding were observed. A typical stress distribution for 7 and 23 fiber RVE without any crack is shown in figure 4.

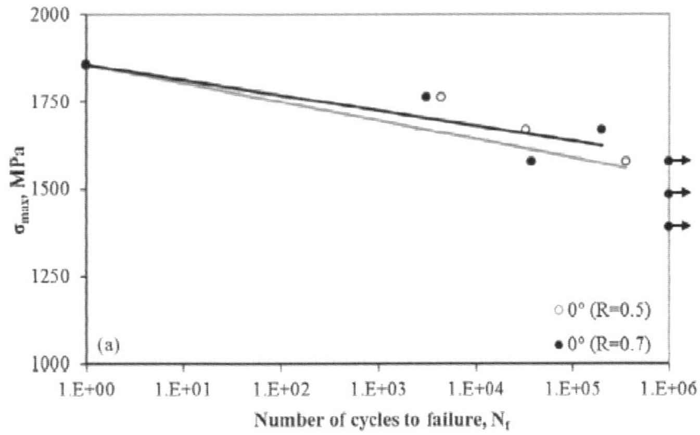

Figure 2. S-N curves for unidirectional laminate, 0^0

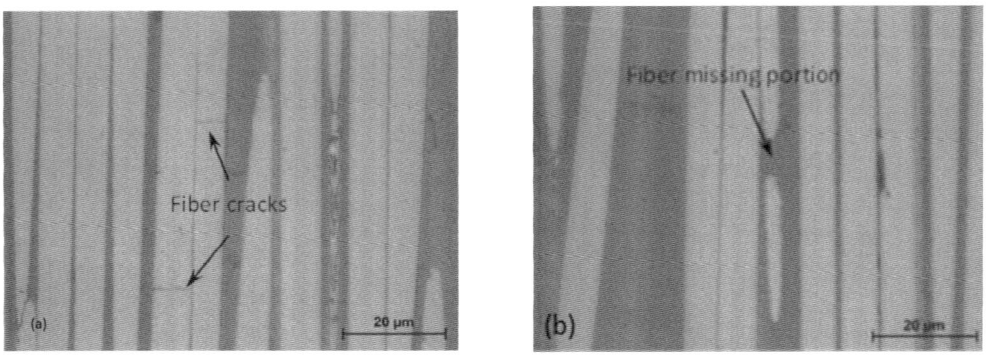

Figure 3. Micrographs (a) unidirectional 0^0 specimen: fiber cracks at center of specimen at plan section, (b) 0^0 specimen: fiber missing portion,

The load was applied by increasing the strain along the longitudinal direction until maximum stress, corresponding to the experimental fatigue test, was reached. Stress distribution for 7 fiber RVE, with crack in center fiber, obtained using finite element analysis is shown in figure 5(a), (b). As expected, axial stress (S11) in the cracked fiber was zero at the crack location. At the crack location, stress from the broken fiber was redistributed to the neighbouring fibres. Stiffness of 7 and 23 fibre RVE composite was not very different and did not change with length of RVE. Hence, it was concluded that the stiffness of composite without any crack was independent of depth of RVE (shown later in figure 6).

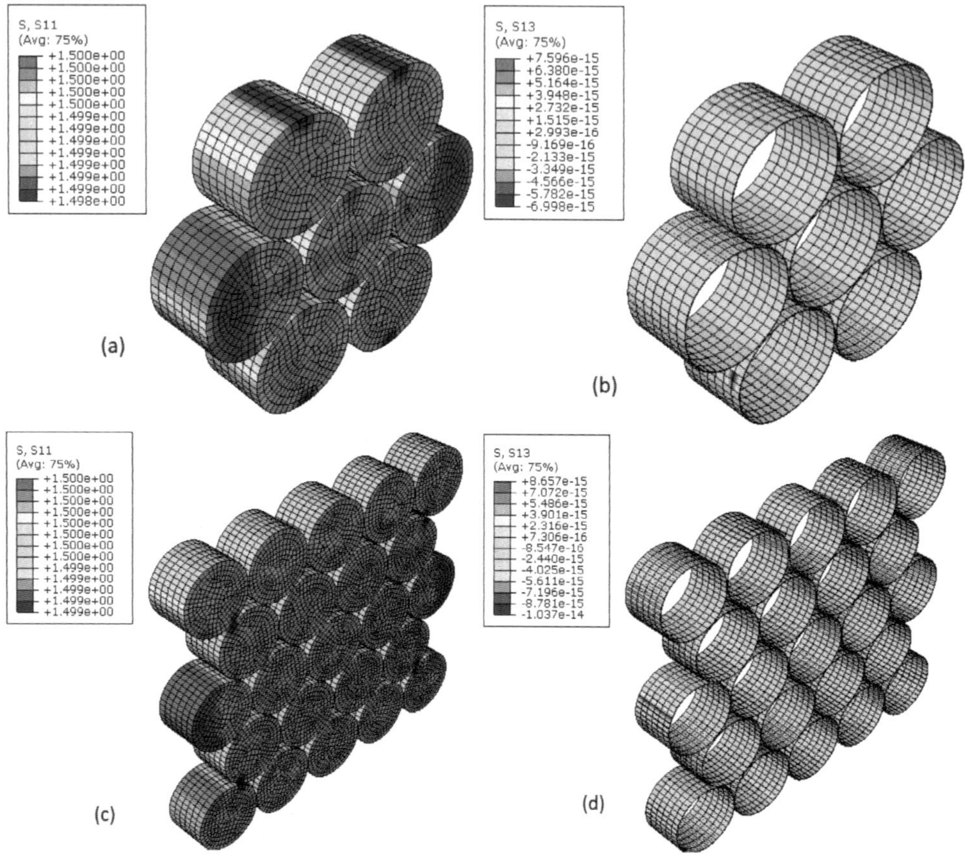

Figure 4. Axial stress (S11, GPa) in the fiber without crack for (a) 7, (b) 23 fiber RVE, shear stress (S13, GPa) in the interface for (c) 7, (d) 23 fiber RVE [0 μm (N_v)]

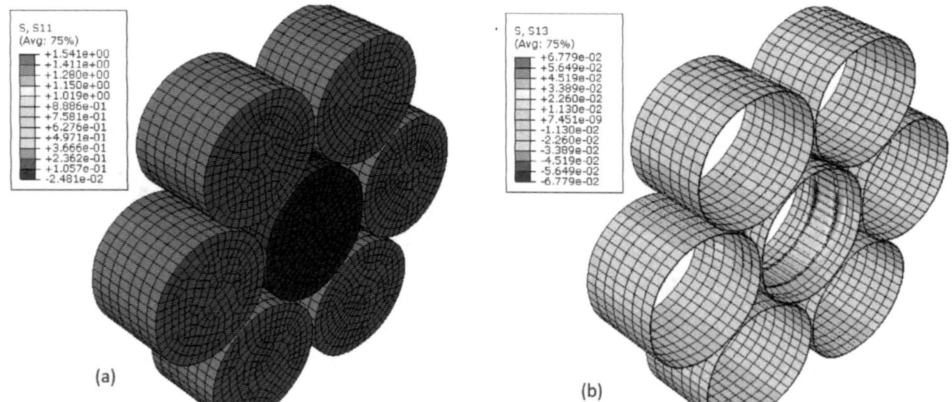

Figure 5. (a) Axial stress (S11, GPa), (b) shear stress (S13, GPa) in the fiber with crack for 7 fiber RVE

Finite element simulations carried out with 7 fiber RVE and 227 μm depth for different fiber crack densities were compared with experimentally measured stiffness as shown in figure 7. It was seen that the stiffness of composite obtained from the finite element simulation was in good agreement with experimental results . However, at R=0.5, there seems to be an outlier that needs further investigation. Thus, a method was established to determine the size of the RVE to predict stiffness reduction in a composite material for a given microstructural damage state and was confirmed with experimental results.

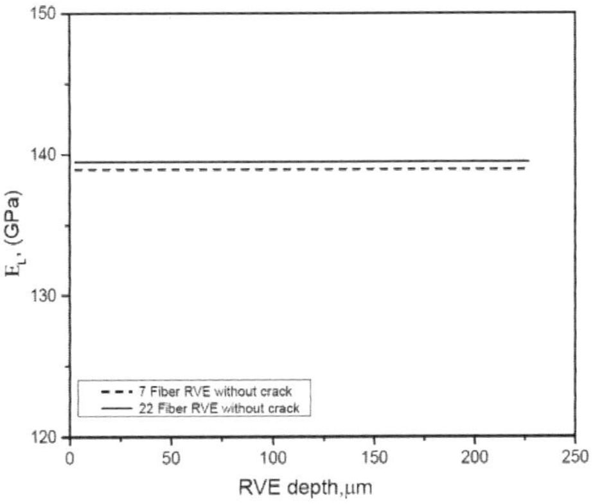

Figure 6. Stiffness of 7 and 23 fiber RVE

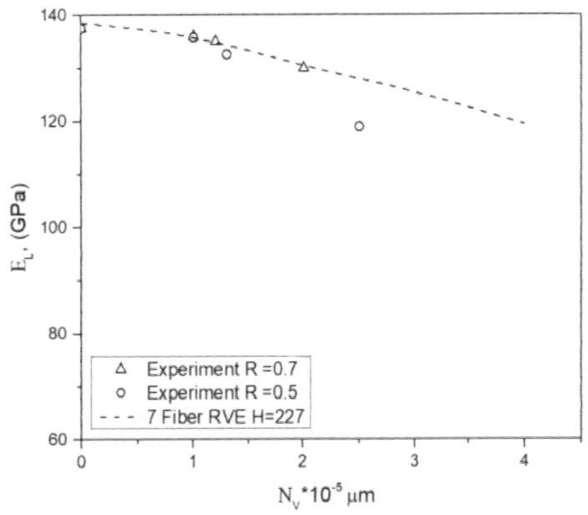

Figure 7. Modulus calculated from FEA

5. Conclusions

Microstructural damage state of a unidirectional CFRP composite was obtained by subjecting it to tension–tension fatigue load for a given number of cycles and stress ratio. Damages were observed in the form of fiber cracking, interface debonding and matrix cracking. Finite element analysis (FEA) was done to develop the methodology to predict the effective stiffness of the CFRP composite for a given damage state. Effective stiffness of composite without crack was found to be independent on RVE depth. Effective stiffness of composite with cracks decreases as the number density of fiber cracks increases. This was found to be in agreement with experimentally observed data.

6. References

[1] Sevenois R, and Van Paepegem W 2015 Fatigue damage modeling techniques for textile composites: review and comparison with unidirectional composite modeling techniques *Applied Mechanics Reviews*, **67** 21401

[2] Degrieck J, and Van Paepegem W 2001 Fatigue damage modeling of fibre-reinforced composite materials: review *Applied Mechanics Reviews*, **54** 279

[3] Harris B, Gathercole N, Lee J, Reiter H, and Adam T 1997 life-prediction for constant-stress fatigue in carbon fibre composites *Philosophical Transactions Mathematical, Physical and Engineering Sciences* **355** pp1259-1294

[4] Quaresimin M, Susmel L and Talreja R 2010 Fatigue behaviour and life assessment of composite laminates under multiaxial loadings *Int J Fatigue* **32** 216

[5] Senthilnathan K, Hiremath C, Naik, Guha A and Tewari A 2017 Microstructural damage dependent stiffness prediction of unidirectional cfrp composite under cyclic loading, *Compos. A*, **100** pp 118–127

[6] Baddeley A, Jensen EBV 2004 *Stereology for statisticians* Chapman and Hall/CRC New York pp48

[7] Melro A, Camanho P, Pires F and Pinho S 2013 Micromechanical analysis of polymer composites reinforced by unidirectional fibres: part ii – micromechanical analyses *International Journal of Solids and Structures* **50** pp 1906-1915

[8] Soden P,Hinton M, and Kaddour A 1998 Lamina properties, lay- up configurations and loading conditions for a range of fibre- reinforced composite laminates *Compos. Sci. Technol* **58** pp 1011–1022

[9] Barbero, E 2008 *Finite Element Analysis of Composite Materials* CRC first ed pp 222-225

Acknowledgments
The authors would also like to thank Senthilnathan, K and Dr. Chandrashekhar P. Hiremath for their support in this research.

AUTHOR INDEX